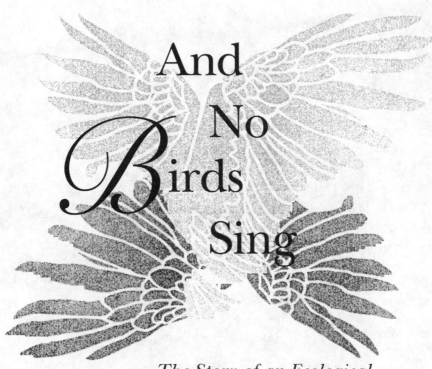

And No Birds Sing

The Story of an Ecological
Disaster in a Tropical Paradise

MARK JAFFE

SIMON & SCHUSTER

New York London

Toronto Sydney

Tokyo Singapore

SIMON & SCHUSTER
Rockefeller Center
1230 Avenue of the Americas
New York, New York 10020

Designed by Pei Loi Koay
Manufactured in the United States of America

10 9 8 7 6 5 4 3 2 1

Library of Congress Cataloging-in-Publication Data
Jaffe, Mark
   And no birds sing: the story of an ecological disaster in a
tropical paradise / Mark Jaffe.
      p.   cm.
   Includes index.
      1. Birds—Guam. 2. Brown tree snake—Guam. 3. Brown tree
snake—Ecology—Guam. 4. Biological invasions—Guam. 5. Pest
introduction—Guam. 6. Birds, Protection of—Guam. 7. Wildlife
reintroduction—Guam. 8. Savidge, Julie. I. Title.
QL694.G8J34   1993
598.25'09861'17—dc20   93-33631
                                                         CIP

ISBN: 0-671-75107-7

Photo Credits
1,5: Anne F. Maben, Guam DAWR; 2,8:Tom Siebert; 3: Gordon Rodda,
U. S. Fish and Wildlife Service; 4: T. H. Fritts, U. S. Fish and Wildlife Service; 6: Gary Wiles, Guam DAWR; 7, 9, 12: Louis Sileo, U. S. Fish and Wildlife
Service; 10: Earl Campbell III; 11: Annette Donner; 13, 14, 16: Stuart L. Pimm;
15: Robert E. Beck Jr., Guam DAWR.

Acknowledgments

It took many steps to come to the end of this journey and I was helped along the way by many people. Indeed, without that help the trip could never have been completed.

If it had not been for the support and fine editing I received from Fran Dauth and Hank Klibanoff of the *Philadelphia Inquirer*'s national desk on my initial Guam stories, I don't believe the journey would have begun. Without the insight and guidance of my agent, David Black, an interesting story would not have been turned into a book.

That book, however, could never have been completed without the cooperation, patience, and support of the scientists who worked on the problem of Guam's disappearing birds.

It was Larry Shelton, the former curator of birds at the Philadelphia Zoo, who first brought this bizarre occurrence to my attention, and he and the zoo continued to help me cover the story over the years. John Groves, the curator of reptiles, and Beth Bahner, the animal collections manager, also provided me with vital information.

On Guam, Rufo Lujan, the director of the Division of Aquatic Wildlife Resources, and his staff were invaluable in providing me with an understanding of the island and its habitat. Once again, special thanks go to Bob Anderson, Gary Wiles, Herman Muna, and Tino Augon.

Acknowledgments

It was with the aid of a small army of biologists that I came to learn about snakes in general and the brown tree snake in particular. Tom Fritts led this band and provided me with much time and insight. His tutorials were augmented by Mike Mc-Coid, Gordon Rodda, and Earl Campbell.

David Chiszar, Keith Kardong, and Robert Mason—three scientists looking at distinct aspects of the snake's biology—also gave generously of their expertise.

When it came to understanding the efforts to save Guam's indigenous birds, Bob Beck was a font of island history and biology. His hospitality and unflagging aid were among my most valuable resources.

Scott Derrickson, of the National Zoo's Conservation and Research Center, helped me understand the nuances of captive breeding, and Susan Haig explained the genetic problems faced by small populations.

The issues of population dynamics, conservation biology, and island extinction were all masterfully addressed for me by Stuart Pimm. Greg Witteman was an able guide through the forests of Rota and provided fine detail and anecdote on efforts to release the Guam rail on that island.

My very special thanks go to Julie Savidge, whose work and documents formed the backbone of much of my story. Time and time again I sought her help and she graciously responded.

Once I had finally amassed this material, a number of people aided me in fashioning it into a cohesive story.

Foremost among those was my wife, Sandra Stuart, an inventive editor and writer. Eileen Brandt served as critical reader for much of the manuscript, and my son Aaron's continued interest assured me that this was a story worth telling.

I must also thank my wife and son for their patience during the work on this book—a period during which I was on the road or squirreled away at the computer much of the time.

Finally, at Simon & Schuster, Dominick Anfuso and Cassie Jones offered astute editing and an attention to detail that I truly appreciated.

For Aaron

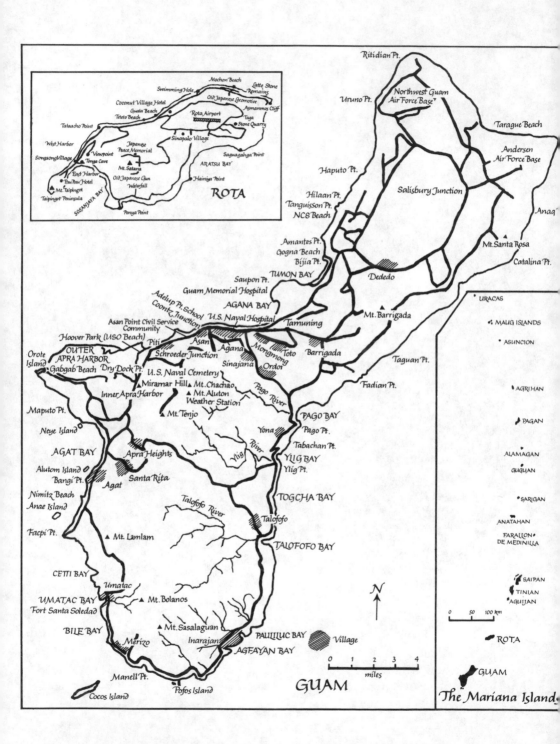

ROTA

Mochon Beach
Latte Stone Remains
Swimming Hole
Old Japanese Locomotive
Asmamos Cliff
Coconut Village Hotel
Guata Beach
Teteto Beach
Tataacho Point
Rota Airport
Taga
Stone Quarry
Sinapalo Village
West Harbor
Japanese Peace Memorial
Saguagahga Point
Songsong Village
Viewpoint
Tonga Cave
Mt. Satana
ARATSU BAY
East Harbor
Pau Pau Hotel
Old Japanese Gun
Waterfall
Hainiya Point
Mt. Taipingot
Taipingot Peninsula
SOSANJAYA BAY
Ponya Point

Ritidian Pt.
Uruno Pt.
Northwest Guam Air Force Base
Tarague Beach
Haputo Pt.
Andersen Air Force Base
Salisbury Junction
Anaq
Hilaan Pt.
Tanguisson Pt.
NCS Beach
Mt. Santa Rosa
Catalina Pt.
Amantes Pt.
Gogna Beach
Bijia Pt.
Dededo
TUMON BAY
Mt. Barrigada
Saupon Pt.
Guam Memorial Hospital
AGANA BAY
U.S. Naval Hospital
Tamuning
Taguan Pt.
Adelup Pt. School
Coontz Junction
Asan Point Civil Service Community
Hoover Park (USO Beach)
Piti
Asan
Agana
Mongmong
Toto
Barrigada
OUTER APRA HARBOR
Orote Island
Schroeder Junction
Sinajana
Ordot
Gabgab Beach
Dry Dock Pt.
U.S. Naval Cemetery
Fadian Pt.
Miramar Hill
Mt. Chachao
Inner Apra Harbor
Mt. Aluton Weather Station
Pago River
Maputo Pt.
Mt. Tenjo
PAGO BAY
Neye Island
Yona
Pago Pt.
AGAT BAY
Tabachan Pt.
Apra Heights
Ylig River
YLIG BAY
Alutom Island
Ylig Pt.
Bangi Pt.
Agat
Santa Rita
TOGCHA BAY
Nimitz Beach
Anae Island
Talofofo River
Talofofo
Facpi Pt.
TALOFOFO BAY
Mt. Lamlam
CETTI BAY
Umatac
UMATAC BAY
Mt. Bolanos
Fort Santa Soledad
BILE BAY
Mt. Sasalaguan
PALILILUC BAY
Village
Merizo
Inarajan
AGFAYAN BAY
Manell Pt.
N
Cocos Island
Pofos Island
GUAM
0 1 2 3 4
miles

URACAS
MAUG ISLANDS
ASUNCION
AGRIHAN
PAGAN
ALAMAGAN
GUGUAN
SARIGAN
ANATAHAN
FARALLON DE MEDINILLA
SAIPAN
TINIAN
AGUIJAN
0 50 100 km
ROTA
GUAM

The Mariana Islands

Preface

A press release came across my desk at the *Philadelphia Inquirer* one day announcing a new display of tropical birds at the Philadelphia Zoo. As the paper's environmental writer I receive such releases all the time.

What was interesting about this one was that Larry Shelton, the zoo's curator of birds, had personally trapped these birds of paradise in the jungles of Papua New Guinea. So I went over to the zoo to hear about how he did it.

The interview was over, our pictures taken, my notebook in my jacket pocket when Shelton said, "You know, if you are looking for a really interesting story, you ought to write about Guam. All the birds are vanishing."

He told me about all he had seen on the island the previous summer on his way back from New Guinea. I was fascinated and expressed interest. But there wasn't much I could do with it at the time.

Nearly a year later, I got a call that Bob Beck, of the Guam Division of Aquatic and Wildlife Resources, and a cargo of Guam birds were arriving at the zoo. I interviewed Beck and then traveled down to the National Zoo's Conservation and Research Center in Front Royal, Virginia, to meet Scott Derrickson, who had become an important member of the team trying to breed in captivity Guam's endangered birds.

My first story on Guam appeared on the front page of the *Inquirer* on June 4, 1985, under the headline "Cracking an ecological murder mystery." I never imagined there would be another. But in late 1989 I heard that there was a plan to return some of the captive birds to the Micronesian wild, and with the support of the *Inquirer*'s national desk I flew out to Guam in January 1990. The result was two more stories.

I suppose work on this book really began that day on a lush plateau of an obscure little island when six Guam rails were set free. It would never have been completed, however, without the cooperation and patience of the many scientists who have worked on this problem.

Over the course of two years, I amassed more than a hundred hours of tape from interviews with the twenty biologists who have probed various aspects of Guam's problem. In addition to their time and recollections, many of them made available to me their scientific papers, their data, and most important, their field notes.

The journey to gather this material took me from New York City to Knoxville, Tennessee, to Lincoln, Nebraska, to Boulder, Colorado, with a number of stops along the way. Ultimately, the story took me back to Guam, where I spent more than a month trying to reconstruct the history contained in field journals and research articles.

Of particular help to me on the island was the Guam Division of Aquatic and Wildlife Resources, which provided me with its extensive records and reports, and the Micronesian Area Research Center at the University of Guam.

The final product is, I hope, more than just a tale of a single island and its birds. For Guam's woes are symbolic of problems that beset many islands and many lands, and the lessons learned there may eventually aid in saving and restoring other wildlife and wild places.

Chapter

1

A fiery sun rose out of the Pacific, turning the dark waters blue and bleaching the sky white. It rose higher and higher, showering bright heat on the hills, forests, and cliffs of a little island.

But below the cliff line the woods remained in the dark, moist shadows. Finally, the sun climbed over the bluffs and its warmth and light poured into the lush crescent of rain forest wedged between the base of the soaring, gray limestone wall and the stainless foam created by coral reefs and the sea.

Here coconut, wild fig, breadfruit, cycad, ironwood, and pisonia trees mingled their fronds, leaves, and needles in a luxuriant curtain of green. Beneath the low canopy, China inkberry bushes and hibiscus covered the ground.

This was a classic "typhoon climax" rain forest—constantly knocked down by the huge storms spawned roughly 3,000 miles to the east in the Pacific "intertropical convergence." It is there, where northeast and southeast trade winds meet, that these storms are born and then sweep west toward Asia, picking up rain and howling winds as they go. The island was right in their path. In some years, as many as thirty typhoons raked the forests, so the trees rarely reached more than forty feet before a really big storm would come and knock them down. As a result, the woods were densely packed, the ground cover thick and verdant.

As the sun peered over the cliff, it warmed these woods and, as if animated by the morning's glow, they became alive with motion and sound.

T*here was the* bree, bree *of the Micronesian kingfisher, a little bird of rust and black. There was the pure two-beat whistle of the glossy, dark blue flycatcher, and the melodious* peter-peter-peter *of the broadbill. High overhead the piercing* caw, caw *of the black Marianas crow could be heard.*

Red cardinal honeyeaters, cinnamon-colored rufous fantails, and flocks of the incredibly tiny, green-tinted bridled white-eyes soared through the woods.

On the ground the glottal sound of the flightless rail—a brown, pigeon-sized bird with the speed of the roadrunner—could be heard, as parents with a covey of chicks scuttled through the brush.

And so, as it had for thousands of years, morning came to the band of forest that wrapped itself around Ritidian Point on the northernmost tip of the island of Guam.

The sky, above the tree line, was already glowing with heat. The sun was rising out of the ocean with the promise, once again, of another blistering day. Overhead, the heavens were still black and marbled with clouds. The cool, damp night air lingered. But this September day in 1983, like so many others on Guam, promised to be a scorcher.

Larry Shelton sat and watched the sun creeping over the trees and into the forest ravine below. He ached the ache of too many hours in jets, too many time zones, and too many days in moldy, steam-bath tropical forests. Now, as he sat on the deck of this cream-colored townhouse, sipping coffee and trying to gather himself from one more seven-hour, red-eye flight, he realized something very strange was unfolding before him.

A few minutes ago, the deep ravine had been just an inky mass, but now, as the sun's rays slowly streamed over the facing hill, a luxuriant wall of tangen-tangen vine peppered by the fronds of cycads and coconut trees emerged. Behind him rose the island's morning sounds. The crow of a booney rooster. The bark of booney dogs. The engine of an early-morning motorist. But before him was only silence.

"Don't you notice something odd?" asked Bob Beck. This was Beck's deck and it was Beck who seven months ago had met

Shelton in Los Angeles and persuaded him to come to Guam.

Yes, Shelton did notice something. Something that he had never seen, no, not seen, something he had never heard before. The dense forest before him was absolutely silent. In virtually every place in the temperate and tropical world the coming of the sun is greeted by the song of birds, but not in this ravine on Guam.

Shelton, the curator of birds at the Philadelphia Zoo, had just been on an expedition to collect birds of paradise in Papua New Guinea, where the jungles were alive with birds—and a good deal more.

One morning, alone in the New Guinea forest, Shelton was removing one of the brilliantly plumed birds from a mist net, when he found himself surrounded by two dozen men painted blue and yellow and carrying stone axes and spears.

The band had come from so deep in the forest that they didn't even respond to the bits of New Guinea pidgin English Shelton tried on them. Still, it wasn't hard to figure out that they wanted the bird.

The basic attitude of most natives on the island, Shelton had learned, was that wildlife was something you either ate or wore, and that was why the bird of paradise's stunning feathers of orange and yellow were so valued. A small, almost delicate man with thinning strawberry blond hair and a matching fringe of beard, Shelton is no Indiana Jones. Still, he wouldn't give up the bird. Edging backward, smiling, and cooing, "Appy nooon"— the New Guinea equivalent of "Good day"—Shelton worked his way out of the clearing and then flew like one of his forest birds through the jungle.

Yet somehow the complete silence that spread before him on this peaceful Guam morning was even more menacing than painted men with stone weapons. By now, the sky above was blue and strewn with huge, unblemished clouds. Still, not a sound emerged from the woods. "There is something seriously wrong here," Shelton thought. Rachel Carson's compelling metaphor of environmental destruction—"silent spring"—came to mind.

"On mornings that had once throbbed with the dawn chorus of robins, catbirds, doves, jays, wrens, and scores of other bird voices there was now no sound; only silence lay over the fields and woods and marshes," Carson had written. Here silent spring—and summer, fall, and winter—was not a symbol but a reality. Yes, there was something seriously wrong. "The silence was so complete that it seemed to be audible," Shelton later wrote, "and so eerie, I felt like shuddering."

Carson's bleak prophecy was powered by the threat of the indiscriminate use of pesticides. But studies had already ruled out pesticides as the problem here. Shelton had gone to Papua New Guinea to rescue birds of paradise from deforestation caused by heavy logging. But that wasn't the problem on Guam either.

Indeed, it was hard to figure out at first glance what the problem could be. Sitting 3,000 miles west of Hawaii and 1,500 miles east of the Philippines, the Micronesian island of Guam hardly seemed a wasteland. Large swatches of the twenty-eight-mile-long isle were covered with forests and grassy savannas. There were limestone cliffs and volcanic hills. The blue Pacific broke on the vibrant reefs filled with fish. Yet clearly something *was* wrong.

While Shelton wrestled with his astonishment, Beck was satisfied with the dramatic effect. He wanted Shelton to be suitably alarmed. But this was nothing new for him. It was a familiar and depressing fact of daily life. Guam's birds were disappearing.

Island wildlife officials had conducted roadside censuses of birds for years. But by 1974 they had abandoned such bird counts for the entire southern half of the island because there simply were no forest birds left.

In the north—in those forests above the Ritidian cliffs—the numbers of birds had increased between 1974 and 1979 and then slowly started to fall. For example, the number of rufous fantails along one road—surveyed consistently over the years—rose from .8 birds per 100 miles in 1974 to 9.3 in 1976, only to fall to 1.1 in 1978 and back to .8 in 1981.

Since late 1982 it had been Beck's job with the Guam Division of Aquatic and Wildlife Resources (DAWR) to figure out how to save the dwindling populations of fantails, bridled white-eyes, Guam broadbills, Micronesian kingfishers, and Guam rails.

It had taken him only a matter of weeks to reach the conclusion that it was impossible to protect the remaining handful of birds in their natural habitat from whatever was out there killing them. That left him with one alternative—trapping and breeding in captivity those species unique to Guam, at least until the menace could be found and removed.

This presented Beck with yet another dilemma, for while he understood something about genetics and ecology, he knew absolutely nothing about bird breeding or aviculture, and that is why he had sought out Shelton, one of the country's foremost breeders of tropical, soft-billed birds.

But even Shelton's expertise could not assure success. None of these birds had ever been bred in captivity, and for some there wasn't even much experience with members of the same general families.

Just two weeks earlier, Beck and Julie Savidge, a fellow DAWR biologist, had captured a rufous fantail and a Guam broadbill. "We had great expectations. We went very fast back to the office, we had it all set up," Beck recounted. They had placed the birds in two small carrying cages and brought them back to the DAWR compound. The birds were placed in tubular wire cages with branches to perch on and plates of mealworms and water at the bottom.

But by the next morning both birds were dead. Dismayed by the episode, Beck decided he would capture no more birds until Shelton came. The fact that the two birds were dead reflected only on his lack of skill, or so he hoped. "I wasn't so worried that I couldn't do it, other than I didn't want to kill a bird," Beck said. "But I was worried that if Larry couldn't do it, we were in big-time trouble. I was told there was nobody better than Larry with these kinds of birds. There was nobody else out there."

Over the next week, a period of torrential rains, Shelton proved his worth. The first thing he did was build new cages. The tubes of rolled wire mesh that Beck had placed the broadbill and fantail in would have been fine, Shelton explained, "if birds flew like helicopters."

The new cages were three feet long, two feet high, and two feet deep, with perches on either end. "Birds need to be able to fly back and forth," Shelton said. "It is their natural behavior, and if you deny them that, you create a lot of stress."

The food also had to be elevated because perching birds simply won't eat like ground foragers. And Shelton also taught Beck to use a small cloth bag or sock to carry the birds in from the field. "Putting them in a small cage," Shelton explained, "just scares the guano out of them."

Shelton wasn't surprised that the broadbill and the fantail had been lost. "You are dealing with small birds with high metabolisms. They start weakening fast," he explained.

Still, as the days passed Shelton could not shake the unnerving silence. "It wasn't a rain forest paradise. But it had a lot of good habitat left. It should have been a place with lots of bird life," he said.

One day in the field, a small flock of white-eyes swooped through the woods. Beck whipped out his binoculars and watched the flock soar into the forest canopy. But Shelton fumbled for his field glasses and by the time he got them out the birds has disappeared. Little did Shelton realize that his lack of dexterity had cost him the last known look anyone would ever have of a white-eye flock.

The day before Shelton left Guam to return to his duties at the Philadelphia Zoo, Beck stood behind a wall of nylon netting in the woods, pursed his lips, and exhaled a clear three-beat call—just like a broadbill. "These birds are very territorial, so when they hear another bird, they're out to defend their turf," he explained. Mistaking Beck for a bird, a male broadbill obligingly flew into the gossamer trap. Within the space of a few hours they had also bagged another flycatcher. Using carrying bags and the new cages, Shelton showed Beck how to handle and care for the birds. Before he got on the plane, he had them eating mealworms. Shelton felt confident that the birds could be bred. The two deaths were nothing to worry about. The folks on Guam just "did not have avicultural experience. That was all," he concluded.

Although disturbed by what he had seen, Shelton left excited by what he considered an absolutely unique effort in aviculture and conservation biology. They were going to save not just a species but the entire suite of native birds from a tropical island. It was a great ambition. It was also a great illusion.

Chapter

2

February 1979

On a late-winter day, Julie Savidge drove the 108 miles from the foothills of the Sierra Nevada to the flat, arid floor of California's Central Valley for the Eleventh Annual Joint Conference of the Western Section of the Wildlife Society and the American Fisheries Society.

She had a lot on her mind as the winter waned—goshawks, South America, Illinois, doctoral dissertations, and her new husband, Tom. But nowhere among this cascade of thoughts would anyone have found the island of Guam.

It is, after all, one of those places that are hard to place. Is it in South America? No, that's Guyana. Africa? No, that's Guinea. Asia? Where is it? Well, if you drew an east–west line between Honolulu and Manila and a south–northwest line between Brisbane, Australia, and Tokyo, just about where the two lines intersected you would find the island of Guam. In other words, in the middle of the Pacific nowhere.

Guam, "where America's day begins." Magellan was the first European to visit the island, sailing into the bay of the little village of Umatac on March 6, 1521. The aboriginal Chamorros, who had arrived about 4,000 years earlier, sailed out to greet the explorer in their praos—outrigger canoes with sails of woven pandanus leaves. They swarmed aboard, invited Magellan ashore, and helped themselves to anything that wasn't bolted down, including a small skiff. Finally, the explorer's crew was forced to use muskets to uninvite the inquisitive and acquisitive natives.

The islanders that the Spanish explorer found were a tall, muscular people who wore hats and sun visors, but no clothes. They lived in sturdy wooden huts with thatched palm roofs. Their families were matrilineal, and several clans lived grouped as a village under a tribal chief.

Both poets and soothsayers, the *makahanas*, were honored by the people and life was simple as the Chamorros fished and gathered the fruit and roots of the island for sustenance.

Magellan, unimpressed with this tropical paradise, sailed off, having named Guam and its neighboring islands *Las Islas de los Ladrones*—The Islands of Thieves. A few years later, the Spanish monarchy bestowed the more dignified and politically appropriate appellation of the "Marianas" on the island chain, in honor of Maria Anna of Austria, the widow of King Philip IV and benefactor of a Jesuit mission on the island. Thus began an era of indifferent Spanish rule and the Chamorros' fitful, sometimes bloody, conversion to Catholicism.

Guam stayed in Spanish hands until June 20, 1898, when with cannons blazing, Captain Henry Glass steamed into San Luis d'Apra Harbor on the U.S. Navy cruiser *Charleston*. Glass expected to meet Spanish gunboats. Instead, a launch bearing the port captain, the port surgeon, and the sole American on the island amiably greeted the naval officer, much as the Chamorros had welcomed Magellan 377 years before. Glass informed them that the Spanish-American War was underway. It was the first news that Guam had heard about the hostilities. But it hardly mattered, for Fort Santiago was out of ammunition. Glass immediately declared the Spaniards prisoners of war and claimed Guam as a possession of the United States. That's what it's been ever since—save for a brief Japanese interlude during World War II.

Valued by the navy and air force as a forward outpost in the nation's Pacific defenses, it is the only American soil west of the international dateline. Thus the sun rises on Guam fourteen hours before it shines on Quoddy Head, Maine—the easternmost spot in the continental United States.

Guam, "where America's day begins." When visitors slip, as

they often do, and say, "When I get back to the U.S.," exasperated Guamanians reprove, "You are in the United States!"

But the geopolitical significance of Guam was not on Julie Savidge's mind as she drove across California's agricultural plains from Cedarville to Redding on State Highway 229.

She and her husband, Tom Seibert, had both been accepted in the doctoral program in ecology at the University of Illinois and her mind was busily turning over dissertation topics.

The two had just spent eight months on a honeymoon trek through South America. Shouldering a couple of big backpacks, they made their way from Colombia—through Ecuador, Peru, Bolivia, and Argentina—all the way to Tierra del Fuego. Then they worked their way up the Pacific Coast, from Chile to Panama.

She had thought she might do a doctoral thesis on some biological problem in South America. But upon returning to the States, Savidge had taken a job as a biologist with the U.S. Forest Service studying the rare and secretive goshawk.

Savidge hiked the Modoc National Forest in the northeastern corner of California in search of hawk nests and territories. "They had been classified as a 'sensitive species' and I was making sure that timber operations were not jeopardizing the birds," she explained.

"For a while I got really excited by these goshawks and I wanted to work on them for my Ph.D.," Savidge said, "but that didn't work out and then I went and heard Mike Wheeler's talk."

Sitting there in a darkened room, at the wildlife conference, looking at slides and maps of an island that until that second she had never given a moment's thought, Julie Savidge found the subject that was to consume the next seven years of her life. Wheeler, a DAWR biologist, presented a paper entitled "Man–Wildlife Relations on Guam." He talked about Magellan and the Chamorros, the Spanish and the deer, pigs, and water buffalo they brought to the island, how wetlands were lost when the U.S. military asphalted them over for runways. And then he mentioned that many species of birds were in decline and no one knew why. To Savidge it was clear that here was a mystery aching for an answer.

Such serendipity had been part Savidge's career in biology

from its very beginning. As a high school senior filling out admissions forms for Colorado State University she came to the question of her intended major. She was stumped. Her dad suggested home economics or perhaps veterinary medicine. She did, after all, like animals. No. She looked down the list of majors. There at the end was zoology. What was zoology? They got out the *World Book* encyclopedia and looked it up.

"Zoology is the study of animals. Zoologists try to answer many questions about animals. For example, they conduct research to determine how animals carry out the activities of their lives. They also study how different species are related to one another and how species have evolved . . . ," the encyclopedia said.

"Many zoologists work in modern laboratories at universities, at research centers, zoos, and museums. Other zoologists do 'field studies' in the outdoors. These studies might be performed in a wildlife refuge, at the North Pole, in the jungle, at sea, or anywhere else animals live.

"Like other scientists, zoologists conduct research by gathering information in an orderly way. Zoologists often begin their research with an observation that arouses different thoughts as to its meaning."

It sounded good. She wrote "zoology." Incredibly, she stuck with it. "I stayed in zoology the whole time. . . . I started taking courses, particularly field-oriented courses—ecology, mammalology, ornithology, herpetology—and I just got more and more excited about the natural world. It was kind of like a whole area that was opening up to me."

Savidge's enthusiasm for biology permeates her open, handsome face, with its dark brown eyes and brown hair pulled back into a pony tail, and emboldens her. And so, after the lecture, she buttonholed Wheeler to find out more about the disappearing birds and whether anyone was working on the problem. "I went up to him and said: 'That sounded really interesting, and I think I'd do a good job on it. How can I get over to Guam?' He told me to contact Bob Anderson."

Although he would rarely handle a bird himself or tramp through the forest, but rather spent his time in a small cubicle of an office, stuffed with papers and chairs on the verge of collapse, perhaps no single person would be as instrumental in pursuing the mystery of Guam's birds as Bob Anderson.

Indeed, without Anderson's arrival on the island in 1978, it is possible that the birds of Guam might have gone extinct without anyone ever knowing why. But in that year, a new DAWR director, Harry Kami, created new positions for two field biologists and a supervisor. Bob Anderson, a sometime graduate student at Oregon State University and technician with the U.S. Forest Service in Corvalis, was hired as the supervisor.

A small, stout man with sandy hair and a well-trimmed mustache, Anderson speaks in a quiet, precise manner that is peppered with official jargon. He appears the consummate bureaucrat, but his official demeanor belies his ecological fervor.

He inherited an office that had done, he explained, "traditional game management—counting pigs, counting deer, counting doves, and regulating hunting. They did have some bird-related studies, but they were mostly survey and inventory type things."

No Westerners ever displayed much interest in Guam's native flora and fauna. The Spanish introduced domestic animals— cattle, water buffalo, pigs—and game animals, such as deer. It is believed that the deer were brought to the island in the early 1770s by the Spanish governor. Within a short time, they were so numerous that during the rutting season when the full moon hung low over the Pacific, the cries of bucks could be heard resounding across the entire island.

After World War II, the U.S. military flew over the battle-scarred islands of Micronesia, including Guam, dropping the seeds of the tangen-tangen, a viny tree or treelike vine, to revegetate the pockmarked islands and prevent soil erosion. The tangen-tangen took root and grew into thickets so dense that they form virtually impenetrable walls of cellulose.

The black francolin and Philippine turtle dove were also brought to the island as game birds, and the black drongo, which had been placed on the neighboring island of Rota for insect control, flew to Guam. It seemed that everyone was more interested in adding plants and animals to Guam than studying what was there in the first place.

Yet this carrying of species from one place to another is

something that, for better or worse, mankind has continually done. It is how corn journeyed from North American Indian villages to French farm fields, how the mongoose went from Asia to Jamaica, and how kudzu got from Japan to the United States.

In some cases, like corn, the transplant has been beneficial. Often, however, even when the intent was worthwhile, the results have been disastrous. The mongoose was brought to Jamaica for snake control but has become a pest in its own right, and kudzu introduced to reduce erosion has become a voracious weed blanketing the American South.

Anderson was, however, interested in Guam's "native" life, and, reviewing the records, it was easy to see that there were problems on the island. "It was clear something was going on but we didn't know what," he said. "We already knew the birds were disappearing but we didn't know why."

Birds, however, weren't the only concern. The island's population of fruit bats had also been steadily declining. While bats may be considered a nuisance in most places, the Marianas fruit bat or flying fox, known as the *fahini* to the Chamorros, was both a popular cultural symbol and a favorite native dish. No wedding banquet, communion feast, or fiesta was complete without *fahini* in coconut milk.

They were so popular, they were being hunted into oblivion. In 1973, all game shooting of the Marianas fruit bat and the sheath-tailed bat was prohibited by Guam law.

But poaching and typhoons continued to decimate the already failing population. In 1978, Mike Wheeler conducted a bat survey and found perhaps 100 fruit bats still alive. He also discovered that imports of frozen bats for Guamanian consumption were soaring, with nearly 25,000 being shipped annually to Guam from the neighboring Marianas islands of Rota, Tinian, and Saipan, as well as from the more distant Micronesian islands of Palau and Yap.

The 1979 DAWR annual report, the first issued under Anderson, warned that the bat faced extinction throughout all of Micronesia if something wasn't done to curb the populace's appetite for *fahini*.

In August 1978, the DAWR proposed that ten native birds and the two fruit bats be placed on the federal endangered species list. These included two birds found nowhere else in the world—the Guam rail and the Micronesian broadbill—and three subspecies unique to the island—the Micronesian king-fisher, the rufous fantail, and the bridled white-eye. The other birds, such as the cardinal honeyeater and the Marianas fruit dove, lived on at least one other island in the Marianas chain.

As part of the listing process, the DAWR was required to state precisely what threat the endangered species faced. But part of the problem was the division didn't know why the animals were vanishing. So, they offered educated guesses. The flightless rail, the DAWR suspected, was being preyed upon by feral dogs and cats as well as monitor lizards and Philippine rat snakes. The broadbill was suffering from a loss of habitat and typhoons. The Marianas dove was being hurt by illegal hunting. The swiftlet was being done in by pesticide poisoning. The truth was there were no data to prove any of it.

The agency also outlined a plan to study the problem and five possible causes of the decline: pesticides, disease, habitat loss, overhunting, and predators. Then Anderson and his staff started to get the word out that Guam was in trouble. It was part of this effort that brought Wheeler to Redding, California, that February day in 1979.

Savidge wrote to Anderson expressing her interest in Guam's avian problems and asking if there was any temporary employment. Anderson wrote back telling her there was nothing at the moment but to keep in touch.

At first, the news of Guam's wildlife problems did not set off any alarm bells with federal wildlife agencies. In fact, the Federal Wildlife Service (FWS) didn't want any work devoted to the issue—until seven unfinished monographs on Guam's fauna were completed.

The monographs were part of the bureaucratic wreckage Anderson had inherited. The FWS had paid for the reports and it wanted them.

"We believe that the wildlife reports should be given the highest priority . . . ," a FWS administrator in Portland, Oregon, wrote to Anderson. "We urge you to restrict all other activities of the wildlife biologists . . . and have them concentrate their efforts on completing [the] management reports."

Anderson wrote back calling such a course "foolhardy," and

after a hastily arranged meeting in Honolulu between Anderson and FWS officials a compromise was struck—work on the endangered wildlife could go on and the studies would be completed as soon as possible. Thirteen years later, the FWS would still be waiting for the last of the reports.

The FWS moved slowly on deciding what it would do about Guam's emerging wildlife problem. It wasn't until late 1980 that the service contacted Anderson and outlined its plan of attack.

"Because of funding difficulties it is necessary that we attach priorities," the FWS explained. Those priorities set the study on pesticides as number one. Then, if money was available, studies and a census of native forest birds should be done. Then came the study of the impact of black francolins as competitors with native birds. This was prompted by the fact that the expansion of the Indian bird's range appeared to coincide with the decline of the native species. Then would come the work on disease. "The predator study would be the lowest priority," the FWS concluded.

Anderson chafed at the federally imposed shopping list. Perhaps it was the most likely approach, but the decision on what was and was not relevant ought to be made on Guam, not thousands of miles away in Portland, Oregon. The DAWR, however, had to depend upon federal money to do the research. "We did not have the freedom with our endangered species. . . . Basically, we were told what to look at," Anderson said.

By this time the problem appeared to be accelerating. "The rate of decline became so obvious. When I arrived in '78, I could go out almost any day and see a Guam rail. I could take you out and show you a Guam rail. By 1980, we couldn't do that. This very rapid decline brought home the problem to us. Something had to be done and quickly," Anderson said.

True to its shopping list, in the summer of 1981 the FWS funded two studies on Guam: an evaluation of the impact of pesticides on the island and a census of the remaining birds.

The two FWS researchers selected to do the work, Chris Grue and John Engbring, were going to need some help. Anderson put in a call to Julie Savidge in Champaign-Urbana. "I called her and told her, 'If you are here we can probably hire you.' But I couldn't promise her anything. She was here in two or three days and we hired her."

Savidge and Tino Augon, an undergraduate biology student at the University of Guam, worked with Engbring censusing Guam's forest birds. Virtually all the remaining birds were to be found on the northern end of the island and most of them were in a sweeping but narrow arc of rain forest beneath the sharp limestone cliffs at Ritidian Point—a singularly inaccessible spot because it was on U.S. military land. Indeed, the path was barred by the Naval Communications Station, a hush-hush submarine listening post.

With military clearance, Savidge, Engbring, and Augon made their way into the Ritidian forest and macheted several straight paths or "transects" through the woods.

Each day they would walk these routes, stopping every 150 meters, listening for eight minutes, and recording every bird call they could identify. Through many repetitions, they would be able to calculate how many birds were out there.

The island had twelve native land birds; one native wetland bird, the Marianas mallard; four breeding sea bird species; and seven non-native species. The natives were primarily forest birds, like the glossy, dark blue flycatcher or the black and cinnamon rufous fantail. The sea birds included boobies, terns, and herons. The black francolin and the Philippine turtle dove were among the prominent non-native species.

The exercise was an eye-opening one for Augon, who had grown up in Chalon Pago, a village in central Guam, where the birds had disappeared years ago. "I never realized how many birds there were on Guam until I did the census," he said. "I was amazed."

Savidge was enthralled with the work and the island. "That summer was really an exciting summer. There was so much to learn and it was all new to me," she said. "It was very apparent that there was a problem, but at that time I wasn't sure at all what was the problem." She spent some time trying to come up with various hypotheses that could explain it.

In one experiment, Savidge put out dummy nests with quail eggs in the Ritidian forest and on the neighboring island of Rota—which was thick with birds—to measure rates of predation.

She found that far more eggs were destroyed on Rota than on Guam, with rats the principal culprits. So it did not seem to her that predators were the problem. At the end of the summer she returned to Illinois with Guam—that hard-to-locate island—as the new center of her universe.

When John Engbring reviewed the census data, he was surprised and troubled. "At the very northern portion of the range there were very good densities for most species, and then as you got farther south they disappeared. The birds simply petered out," he said.

What also made it confusing was that all kinds of birds were disappearing. Little white-eyes that nested on the highest branches, bigger kingfishers that made their homes in tree cavities, big, gregarious rails that lived on the forest floor—all were vanishing. Birds that ate seeds were disappearing. Birds that ate fruit were disappearing. Birds that ate lizards or insects were disappearing, too.

"My impression was that there was something very seriously wrong," Engbring said. And whatever was the problem, he knew it was unparalleled in the Pacific. Tinian—a tiny Marianas island 130 miles north of Guam, which is best known, if it is known at all, as the launching site for the World War II atom bomb attacks on Japan—had been the scene of fierce fighting during the war. "Tinian was almost obliterated. Photos after the war show this white limestone rock with a solitary tree here or there," Engbring said.

Yet, somehow, a tiny bird called the Tinian monarch made it through the fighting and the defoliation and survived to become abundant and comfortable in the scrubby tangen-tangen planted by the U.S. military. Whatever was out there in Guam's forest, Engbring suspected, was more deadly to the birds than war.

The prime suspect was pesticides. For several years after World War II, DDT was regularly sprayed over Guam by U.S. military planes, and even in 1975, the body tissue of swiftlets showed traces of the deadly chemical.

That same year, an outbreak of dengue fever among thousands of Vietnamese refugees on the island forced the military to spray a third of Guam with Malathion, leading to massive fish kills in several bays. The spraying also coincided with the disappearance of the nightingale reed warbler from the Agana swamp.

Throughout the 1970s, the military continued to use Malathion and the herbicide Diuron 80 for pest control around its installations.

And so, while Engbring was counting birds, Chris Grue collected samples of soil, lizards, shrews, insects, fish, and guano searching for telltale traces of pesticides. Grue was looking at the likely pesticide "repositories," places where traces of the chemicals would build up. Then he compared those levels with the concentrations known to affect song birds and other birds.

"All we found were background levels," he said. An examination of layers of bird guano, collected from rocks, indicated that in the past birds had traces of DDT in their systems but did not appear to have them now.

Furthermore, when he checked pesticide use on the island, he discovered that more than 84 percent of the chemicals employed were organophosphates and carbamates, which break down rapidly in the environment. There simply was no sign that birds were dying because of the application of pesticides on Guam. There might be a silent spring coming to Guam, but it wasn't like the one Rachel Carson prophesied.

"I left the island convinced the problem was disease," Grue said. Engbring agreed, "At that point, it seemed the most likely candidate."

It must be a disease. It made sense. Hawaii's forest birds had struggled for years with avian malaria. Wake Island's albatrosses had suffered from avian pox. And the number of mosquito species on Guam had risen, as a result of increased air traffic, from five before World War II to thirty-five after the Vietnam War. Disease was just the kind of thing that wreaked such havoc and decline among island bird communities.

And so, in late 1982, the DAWR decided that someone was to be hired to find Guam's disease and Bob Anderson again called Julie Savidge.

"Julie impressed me, because I said, 'Well, if you are here, we can probably hire you,' and she was here. Left her husband and everything and came on out," Anderson said. Savidge was happy to get the job. Anderson was happy to have her.

Still, what was happening out there in the forest was moving too quickly to sit and wait for it to be revealed. Anderson felt that something would also have to be done to try to protect the birds that remained. Another job would have to be created for somebody to work exclusively with these failing species.

Ever since he was twelve years old and watched the birds perching on the blossoming cherry tree outside his bedroom window in Hagerstown, Maryland, Bob Beck had been an avid birder, and it was through the "old birder" network that word of Anderson's job reached him. "I got to know the folks at DAWR because we'd gone birding a bit and when this position came up Bob Anderson asked me if I was interested."

And so, in the fall of 1982, Savidge and Beck joined the DAWR to deal, each in their own way, with Guam's vanishing birds.

October 1982

*L*esson number one: Islands are separate not only in space but also in time. All those zones, datelines, clocks, and calendars that are supposed to link the world are meaningless. Guam, for example, runs on "Guam time."

Savidge had been hired, according to the DAWR job description, "to investigate the role of disease" in the decline of the island's avifauna. So, before heading for Guam, she had stopped at the federal Wildlife Health Research Center in Madison, Wisconsin. There she learned how to make blood smears and take viral and bacterial samples from birds and how to pack a bird carcass for the 8,000-mile trip from Guam to Wisconsin. At the Madison laboratory, the carcasses would undergo necropsies—the animal world's equivalent of autopsies.

The best telltale clues for disease, however, would most likely be in the blood. So Savidge was drilled not only on making and staining blood smears but on how to search for parasites, bacteria, and any other hints of disease with a microscope.

But upon arriving on Guam, she discovered that there was no stain. Maybe that wasn't such a big thing, because there weren't any glass slides either.

One of the ways she was going to obtain the birds for the blood samples to smear on the slides was by trapping them in fine meshes of nylon called mist nets. But she found that the mesh of the nets available at the DAWR was so large that the little white-eyes flew right through them.

No stain, no slides, no birds. Oh yes, if she ever got the birds, slides, and stain, she would also require a good microscope. Some crucial tools were missing and there wasn't much that could be done in searching for the mysterious Guam disease without them. So Savidge wanted them right away. That's when she learned about Guam time.

Every island seems to have its own time, and the smaller and more remote the island the slower the time moves. There's Hawaiian time, which means that if a meeting is schedule for 2 P.M., it will probably happen that afternoon. On Guam, that meeting might happen that day or the next. On Yap, 520 miles to the southwest of Guam, they still use stone money.

Perhaps there is no rush because everybody knows no one is going anywhere. Sooner or later all paths will cross. No man is an island on an island. Guam time. Get to know it. Savidge had no choice.

"I had to fill out a purchase order and then it seemed I had to get everybody on Guam to sign it," she said. Weeks passed. Months passed. "It took so gosh darn long," she said. "Julie," an old island hand tried to explain to her, "Guam's a lot like Mexico, except without the sense of urgency." Right. If the Mexican slogan is "Mañana," the Guamanian motto could be "Whenever."

Still, biologists, back to Charles Darwin, have always loved islands. It was the flora and fauna of the Galapagos Islands, some 600 miles off the coast of South America, that provided much of the fuel for Darwin's theories on natural selection.

"The archipelago is a little world within itself, or rather a satellite attached to America," Darwin wrote in *The Voyage of the Beagle*. He marveled that despite the geological newness of the nineteen volcanic isles, there were a multitude of animal and plant colonists. This is how the world must have gone from rock to life. "Hence," Darwin wrote, "both in time and space, we seem to be brought somewhat near to that great fact—the mystery of mysteries—the first appearance of new beings on this earth."

Since then, biologists have returned again and again to islands. The musings of some of the most distinguished researchers have been focused on these dollops dotting the
oceans.

There is good reason. One of the techniques of science is to
reduce the world to simplified models and track variables
within those models. But in the main this is difficult to do on
continents, where the vast ebb and flow of life washes over the
land. Islands, on the other hand, afford discrete and simplified
fields of observation.

Edward O. Wilson, in his *The Diversity of Life,* observed that
"archipelagoes . . . not only are isolated but also small enough
and young enough, in comparison to continents and oceans, to
keep patterns . . . simple and hence decipherable." Islands have
been particularly helpful in perusing questions of evolution,
population, biodiversity, and extinction.

When the finch managed to make it from the mainland to
the Galapagos, it didn't remain the same old finch. Over time,
as the birds spread, they moved into different ecological niches
on different islands, becoming fourteen distinct species and
buttressing Darwin's belief that "species are not immutable."

In dealing with population and biodiversity, islands, with
their clear borders, limited traffic, and more easily quantified
resources, once again offer an ideal biological laboratory. In
1963, Edward O. Wilson and Robert MacArthur developed a
model to explain variations in the number of species on an island. They had noted a strong relationship between the size of
an island and the number of species living there.

The bigger the island, the more species. They found an average of about fifty species on a 386-square-mile island. But for
every tenfold increase in land, the number of species, Wilson
and MacArthur calculated, roughly doubled. This they called
the "area effect."

The location of the island also influenced the number and
distribution of species. The farther the island from continents
and other islands, the fewer species living on it. This they called
the "distance effect."

The islands of greatest interest to biologists have been those
created far out at sea by volcanic action as opposed to former

bits of continents. New Caledonia and New Guinea, for example, were once part of the Australia continent and as a consequence share plants, animals, and natural history with it.

The only living things reaching the newborn oceanic islands are migrants that arrived on the winds and waves. The closer to other land masses, the more immigrants. The bigger the land the more species that can be accommodated.

"The fortunate colonists," Wilson wrote, "originated in large, crowded fauna and flora, pressed by competition, predators, and disease, restricted in habitat and diet. They arrived in a new, mostly empty world where, initially at least, opportunity was spread before them in abundance."

And so unique little ecosystems are created, island by island, with wonderful variations on the basic melody of life. It is believed that several thousand endemic species were created in Hawaii from just 400 immigrants. In general, oceanic islands may have fewer total species than a comparable continental land mass, but much of what they've got will be found nowhere else.

But if islands represent opportunity, they also offer risk. MacArthur and Wilson calculated that the elements of size and distance would also work inversely. A small, distant island would have fewer species, smaller populations, and therefore a greater likelihood of losing its species.

The forces at work here are greater exposure to the elements, limits in population size because of limited resources, and fewer options for escape. As a consequence, island species, in general, face a much greater risk of disappearing. Thus, islands are a good laboratory not only for evolution but for extinction as well.

Until the oceans began rising 10,000 years ago, Tobago, Margarita, Coiba, and Trinidad were all attached to South America and shared the same flora and fauna.

Jared Diamond and John Terborgh studied the disappearance of birds from these "land-bridge" islands and found that the smaller the island the higher the "rate of decay" of species.

In the last 400 years, a total of 724 known species of plants and animals have gone extinct, according to the International Union for Conservation of Nature (ICUN). More than 48 percent lived on islands, even though the vast majority of species are continental.

The story is even more pronounced for birds. While less than 20 percent of the world's bird species are island dwellers, islands have been the scene of 93 percent of the recorded avian extinctions—163 species and subspecies. Now, it seemed that Guam's birds were poised to be added to this list.

Although islands may be natural laboratories for the study of extinction, there is nothing natural about the wave of annihilations sweeping across them in modern times.

"Until the present century, the remoteness and inaccessibility of most of the world's half-million islands served to effectively isolate them as sanctuaries for animals and plants," observed Donald Merton. No longer. Today, by ship and airplane, there is virtually no inhabited island that is not touched by the outside world.

And with each contact, fragile island ecosystems run the risk of being exposed to what Edward O. Wilson has described as "the four horsemen of environmental apocalypse"—disease, predators, habitat destruction, and hunting. So ill-prepared are island dwellers that species still common on continents disappear in the face of the horsemen.

Julie Savidge suspected that she was dealing with one of the apocalyptic riders, and she was ready to take her place in that long line of biologists, from Darwin to Wilson, who have studied islands—if she could only get somebody to sign a purchase order.

Fortunately, she had other things to keep her busy. The DAWR and the U.S. Fish and Wildlife Service might have wanted her to find the mysterious Guam disease, but the Department of Ecology, Evolution, and Ethology at the University of Illinois wanted more.

As part of her doctorate, Savidge would have to demonstrate that other likely suspects were not causing the decline of Guam's birds. In writing up her project prospectus, she had listed five possible causes for the decline: overhunting, habitat loss, pesticides, predation, and disease. To satisfy the university she would have to clear or convict each suspect.

Now she had ample opportunity, while she waited for slides, stain, and nets, to scrutinize some of the other possible offenders. Thanks to Chris Grue's work the year before, pesticides could easily be scratched from the list.

Similarly, overhunting was readily disproved. Hunting was severely restricted on most U.S. military land and still these guarded forests provided no refuge for Guam's beleaguered birds. Next came habitat loss. Savidge assembled maps and aerial photos of the island and plotted all the good habitat and all the populations of birds.

There are two kinds of oceanic islands in the Pacific. Some are limestone platters, where a volcano first pushed up from the sea floor only high enough to sustain reefs and then at some later time, belching and spewing lava anew, raised the entire reef, creating a limestone deck. Sometimes the volcanic activity raised the old reef high enough to create towering cliffs, but often these atolls are barely above the lap of the waves and thus are known as "low islands."

Then there is the volcano that pushed itself up from the sea floor to create an island of rich soils and sharp peaks. These are called "high islands." Guam is both.

From the air, it looks like an elongated footprint in the blue Pacific, with the heel at the northeast and the toes pointing southwest. The heel was formed when an old volcano, already topped with limestone, was lifted out of the ocean by a younger volcano, full of vigor and lava, that popped up out of the ocean to the south.

Micronesia sits at the edge of one of the most active volcanic areas on Earth—the so-called Ring of Fire—and as a result lots of islands have broken through the Micronesian expanse of the Pacific, which covers an area roughly comparable to the U.S. mainland.

There are thousands of islands in four great archipelagoes dotting this watery domain. Only 125 are inhabited, and of those, Guam, thanks to its dual volcanoes, is by far the largest, stretching more than twenty-eight miles.

The result is that northern Guam is a land where sharp limestone cliffs rise nearly 600 feet out of the Pacific. It is covered with viny, gnarled forests of ironwood and yoga trees. There are no rivers, streams, or ponds, as is the case on these limestone plateaus. The southern part of the island, on the other hand, consists of volcanic hills, rippling with vibrant green foliage and bright red clay soils. Mount Lamlam, 1,334 feet, is the island's tallest peak. Although much of the south has been cut, leaving only sword-grass savannas behind, there are still ravine jungles of bamboo, where swift little rivers turn into crashing waterfalls.

Savidge pieced together aerial photos of the entire island, marking the forests that had been lost and the forests that remained. On top of that she overlaid the census of the island's birds.

She found that the distribution of good forest habitat and birds simply didn't match. There were prime forests dotting the island that simply had no birds in them. "It's hard to say this is not a problem, but there didn't seem to be any correlation between loss of habitat and the decline in birds," she said. "That left two possibilities—predators and disease."

There was a nefarious melange of candidates when it came to possible predators. There were dogs and cats that had slipped the hold of domesticity and roved the island in feral packs. There were rats, which had been documented as a problem for birds on other islands. There was a three-foot-long monitor lizard, with sharp teeth, carrion-tainted breath, and a passion for bird eggs. And there was a serpent, widely known as the Philippine rat snake.

Somehow, the snake got onto the island after World War II. Some said that the U.S. soldiers actually brought it to Guam to control rats. Others said it must have come in with all the war materiel being shipped back to the States after the fighting was over. Anyway, there it was, a nocturnal, tree-dwelling reptile that was clearly in a position to eat birds.

Still, there were problems with all the suspects. It seemed far-fetched that dogs or even cats could penetrate the dense rain forests and manage to wipe out all bird species so efficiently. The monitor had lived on the island for thousands of years, apparently in some natural balance with the island's birds. Why would it suddenly tip the scales?

Rats were a possibility. *Rattus rattus*—the black rat—sailed with Columbus (a rat skull was found in one of the explorer's ships that sank off the north coast of Haiti) and they have been causing havoc on islands ever since. Rats have been implicated in more than half of the extinctions of island birds—particularly in New Zealand and Hawaii. Black rats, for example, made their way to New Zealand's South Cape Island, around 1964, and soon caused the extinction of the Stewart Island snipe and Stead's bush wren. And it isn't only birds that they have gone after—the Virgin Islands' tree boa is now an endangered species because of the black rat.

Other mammals have caused similar destruction on other islands. Cats decimated a species of dove that lived on Socorro Island, a little Mexican possession in the Pacific. Mongeese have caused the extinction of the Jamaican least pauraque and Hawaii's dark-rumped petrel. But rats have been on Guam for more than a century. Why would they suddenly start wiping out the birds now?

The rat snake seemed even less plausible. Being a slow-moving animal with a low metabolism that may eat a single meal in a month, snakes simply weren't associated with the kind of ecological destruction wrought by active and hungry mammals, such as rats, cats, and mongeese. Besides it seemed unlikely that the snake could be preying on all the species, from the tiny white-eyes to the big, aggressive, ground-dwelling rails. There wasn't a single record of a snake creating such problems on an island.

Still, the snake was a relative newcomer to the island and was a curious animal. Although many people called it the Philippine rat snake, Savidge knew it was actually the brown tree snake and it came not from the Philippines but from the islands of the western South Pacific.

Drab, olive brown in color, with a light, almost luminescent underbelly, it was thin as the shaft of an arrow and had a slender, elliptical head with jaws filled with needle-thin, needle-sharp teeth. While most of the snakes grew to be no more than a few feet long, there had been some specimens captured that were eight to eleven feet in length.

A predator of the night, it moved through the trees and was

difficult to spot, owing to its natural camouflage. It undoubt-
edly preyed on birds, but virtually all the islands in its native
range had birds.

Savidge started digging through old DAWR records looking
for references to the predators. She checked the library records
at the University of Guam and the morgue of the *Pacific Daily
News*. At the DAWR offices she found stacks of field observations by DAWR biologists, on three-by-five file cards, stretching
back to the 1960s. The remarks were arranged by year and
species and would read like this: "I found a snake in Dededo.
Opened it up; found nothing inside."

Acting more like a social scientist than a biologist, Savidge
also went to Guam's people. Savidge devised a questionnaire to
try to pinpoint when people last remembered seeing birds in
their villages and what problems they might have had with disease, rats, snakes, or other predators on their poultry and domestic birds. "All I knew was that I had a different sort of
situation than a lot of scientists have, and I knew that I had to
get as much data as I could and I had to get it any way I could,"
Savidge said. "As the project went on I felt more like a detective
than a scientist."

She published the questionnaire in the local paper and asked
people to send them in to the DAWR. Herman Muna, a young
Chamorro recently hired by the DAWR as a technician, was also
pressed into service as Savidge's pollster.

Muna would drive to a town like Inarajan, a neat little farming village on the southern coast, where the main street runs
parallel to a blue Pacific bay and ends at an imposing white concrete church. There he would knock on door after door from
the bay to the church, asking the dozen survey questions of anyone who would cooperate.

Just outside of Inarajan, Clay Carlson and Patti Jo Hoff had a
small ranch. It was more a hobby and a getaway spot than an
economic concern for the two University of Guam professors.
Still, they were earnestly trying to raise pure-bred rabbits,
geese, ducks, and chickens. But they were having little success.

"We went two years without hatching anything," Carlson said.

For a while, Carlson and Hoff were baffled, but eventually their problem became clear. "The snakes were getting everything," Carlson said. Thus began a battle between the would-be ranchers and the serpent.

"We tried to snakeproof the cages, but they seemed to be able to get into everything," Carlson said. "They even got through a half-inch wire mesh."

The couple took to collecting eggs and hatching them in a mechanical incubator and building cages within cages, using a fine metal mesh to protect fledglings. "We had to take all small animals from their mothers and rear them, to protect them," Hoff said.

Still, every time eggs or young animals were around so was the snake. During one eight-month period, Carlson captured forty-three snakes on his property. "They're mean and they bite," he said.

Even with these precautions it was difficult to raise animals. On several occasions eggs were lost in the incubator because of power outages, outages caused by snakes hitting power lines. In fact, 1982 was a particularly bad year, with sixty-five snake-induced blackouts—a record number.

"We've been trying to come out even on this ranch, but we haven't even come close," Hoff said. At least, it was a hobby for the couple. "If you were a poor farmer," she observed, "and you were struggling with this problem, you'd be screwed."

As such reports came to Savidge she realized that the brown tree snake was an increasingly likely suspect. There were too many complaints from people like Carlson and Hoff and too many blackouts. (Savidge even began collecting statistics on the power outages.) The snake must have some role in what was happening on Guam.

Still, the idea that a snake was wiping out *all* of Guam's birds was hard to believe. If the snake was the cause, it was an unprecedented chapter in natural history, and this meant that Savidge would have to find some unique way to prove it.

One thing was certain: the birds were continuing to disappear. When Savidge had worked with Engbring in 1981, she had found good populations of birds at Ritidian Point and just to the south on U.S. Air Force land at Northwest Field and Andersen Air Force Base. Now, a little more than a year later, there were only a handful of birds left in the forests that grew between the old runways of the airfield. "There were some but not many," she said.

4

February 1983

*L*arry Shelton was at a banquet at the Sheraton Premier Hotel in North Hollywood when he looked up to see Guy Greenwell ushering in a tall, gangling figure with angular features, framed by tousled black hair and a trim black beard. Greenwell, the curator of birds at the National Zoo in Washington, D.C., introduced this Lincolnesque figure as Bob Beck.

Beck, Shelton, and Greenwell had all come to the Sheraton for the Third Annual Jean Delacour Symposium, a conference on aviculture sponsored by the International Foundation for the Conservation of Birds. The foundation, also based in North Hollywood, was conceived by Gerald L. Schulman, a Los Angeles real estate promoter, philanthropist, and bird enthusiast. Schulman bankrolled the foundation, the symposium, and the grants the foundation gave for research into issues such as the comparative behavior of penguins and the natural incubation of parrot eggs. He had also donated $100,000 to the San Diego Zoo for the construction of an exotic bird hatchery in his name.

Schulman had made his fortune investing the money of the movie-industry affluent—including Robin Williams and Woody Allen—and numerous professional athletes in a real estate syndicate that purchased 600 properties in forty-seven states. Two-thirds of those buildings were used as U.S. post offices and generated $40 million a year in rent.

And so, the cream of the bird community gathered under Schulman's largess to listen to papers with titles like "The Breeding of Hartlaub's Duck" and "Can the Socorro Dove and Socorro Mockingbird Be Saved?"

Unfortunately, for the aviculture community, it turned out that Schulman had been convicted of business scam felonies twice on the East Coast, before heading west. Actually what was really unfortunate was he hadn't learned his lesson.

In February 1988, he was convicted again, this time for tax fraud involving dummy loans from Panamanian and Dutch Antilles banks. The Jean Delacour Symposium and the International Foundation for the Conservation of Birds became victims of the FBI and the U.S. Attorney's Office.

But when Beck arrived, on February 24, 1983, for the three-day symposium, the foundation was at its zenith and the Sheraton was packed with aviculturists, zoo curators, and biologists. Beck did not know what to expect. "I went to the symposium completely cold," he said.

Soon after being hired by the DAWR, Beck launched a series of field studies to determine how many flycatchers, white-eyes, fantails, and rails were left.

While Beck hiked the forests of Guam, Tino Augon, who had parlayed his summer job censusing birds into steady employment at the DAWR, sat down at division headquarters and gathered about six years of the bird census. These counts had been done by habit and rote, but Augon decided to see if the numbers revealed anything more than the steady disappearance of the island's birds.

Like Bob Anderson, Augon had become a fixture at the white stucco building that housed the DAWR biologists. Now, he was an old hand. He had seen biologists come and go. "Everybody had done a little work on the problem, but nobody had an answer. There was a real feeling of helplessness," he said.

Using the old surveys and a personal computer, Augon discovered that each surveyed area appeared to be losing about two species each year, with the smallest birds apparently going first. In fact, he calculated a "susceptibility index." The bridled white-eye, the broadbill, and the rail had the greatest vulnerability, the Micronesian starling, the Micronesian kingfisher, and the Marianas crow the least.

Augon calculated that if whatever was decimating the birds hit the remaining populations at Ritidian and Northwest Field all the forest birds would be gone in five years. Beck feared that this last bastion might go even faster, since the other sites might have been bolstered as some birds migrated in from the more besieged areas of the island.

"Tino's work basically showed us we didn't have a lot of time," Beck said. There simply was no way to preserve the birds in the wild. The only chance, Beck concluded, lay in captive breeding the island's key endemic species—the rail, the white-eye, the fantail, the kingfisher, and the broadbill.

The problem was no one on Guam knew anything about raising wild tropical forest birds. Beck was going to need help. This did not faze him. What he did not know he would learn. What he could not do himself, he would find those who could.

He was now in his late thirties, and his beard was revealing just a hint of gray, but Beck still described himself as "a child of the sixties." Yes, he did have a taste for tofu, wire-rimmed glasses, and mountain bikes, but his laid-back nature masked the zeal that sent a generation into the streets to protest.

Guam's birds might be in peril, but they had gained a stalwart advocate who would do battle with the federal bureaucracy, the military, or anyone else who might stand in the way of rescue.

Beck sat down and wrote a letter to S. Dillon Ripley, the secretary of the Smithsonian Institution and one of the most influential people in American science. Ripley had, it also happens, written the definitive book on rails.

The letter laid out Guam's problem and asked for assistance. "He wrote a letter back saying, 'I think we can captive breed the rail fairly easily and the National Zoo will be able to be of assistance to you,' " Beck said.

More letters followed to the Bronx and San Diego zoos and wildlife officials in Australia and New Zealand, who were already engaged in captive breeding exercises.

Then Beck heard about the Jean Delacour Symposium. He

asked Anderson whether he thought it was worth the 6,000-mile trip to check it out. Anderson said, "Go."

Beck's aim at the symposium was to interest mainland zoos in helping the DAWR. It turned out not to be a difficult exercise. In fact, Beck found that curators from all over the country were seeking him out to hear about Guam's strange problem.

"Well, it turns out that Dillon Ripley had turned all these people on to us, like Don Brunning from the Bronx Zoo in New York City. Guy Greenwell, who was then the dean of American aviculturists, was there too," Beck said. It was Greenwell who took Beck in tow and began introducing him to people he thought might be useful, people like Shelton. Beck and Shelton arranged to meet and discuss Guam's problem. When the story was laid before him, Shelton was astounded and confounded.

Within a day, Beck had become a conference celebrity. "All these folks were coming up and wanting to hear the story and what was going on and wanting to help," he said. Curators from zoos in Dallas, Denver, Fort Worth, Honolulu, San Diego, and St. Louis all told him that they might participate in some program to save Guam's birds.

Beck was riding a wave of serendipity that seemed to have stretched over the years and all the way from Kingston, Rhode Island, to Guam. He had left the University of Rhode Island, where he had been a graduate student in evolutionary genetics, with no firm plans. He had taught high school in Boulder, Colorado, and when his contract wasn't renewed, he got another contract teaching on Guam.

He had left Guam, in 1978, to work on a Ph.D. in ecology—spending three years in the Smoky Mountains of Tennessee studying vireos—and had returned to the island with the intention of writing his dissertation.

But then along came his birding buddies with this job working with endangered species. The interesting part of the doctorate—the field work—was over. Now, he faced taking his storklike, six-foot-two-inch frame and putting it behind a word processor and staying there and writing. But how could typing a dissertation compare with the "mission" of saving endangered species? The thesis would always be there waiting—the birds would not. Now, the entire American zoo world seemed to be at his service.

In fact, it appeared that events were now out of Beck's hands. One day in the hotel lobby, Shelton, Brunning, and Greenwell decided that there ought to be a Guam bird rescue project. Birds would be trapped and brought to zoos in the United States. "The three zoos were collaborating in something we called 'Zoo East' and it was just the kind of project that seemed right for us," Brunning explained. Beck had managed to enlist the resources of three of the country's major zoos before he even knew it.

They were all exhilarated by the prospect of such a sweeping program. "It was the first time that an entire avifauna was the target," Shelton explained. "There had never been an attempt to save an island's entire endemic avifauna."

The initial desire was to bring in the small passerines—like the bridled white-eyes and the fantails—which Augon's work had shown were at the greatest risk.

This was easier said than done. While white-eyes have successfully been bred, there was much less experience with comparable broadbills and fantails. Beck was arguing that all the birds should be brought in, but at that time there wasn't a panic about the rail or Micronesian kingfisher. A rail program was also cobbled together, however, because it was a gregarious and fecund bird that the breeders thought would be easy to work with and it was incidentally the species Ripley had the most interest in.

It was clear to the band of curators from the very beginning that this would be a complex and expensive project that would need the cooperation of many zoos, which would have to be willing to commit not only space and expertise but money as well, for there was little prospect of getting money from either the federal or island governments.

Although Beck did not realize it at the time, he and Guam were about to become the beneficiaries of a fresh revolution in thinking in the American zoo community. It had been less than three years since the American Association of Zoological Parks and Aquariums (AAZPA), at its midyear meeting in Tulsa, Oklahoma, voted to make wildlife conservation its highest priority and captive breeding its major tool.

Indeed, until that step was taken, zoos had traditionally been consumers of wildlife, not saviors. But by the late 1970s, many curators and biologists realized that their stores in the wild were running perilously low. As a result, the AAZPA decided to create "species survival plans," to raise and manage populations of key endangered animals in captivity.

Central to the idea would be efforts to safeguard the genetic viability and variability of the captive population. This, too, was a new idea for the zoos. For although any animal breeder or chronicler of European royalty could tell you that inbreeding is a bad idea, it was common practice and hardly a concern in the zoo community.

In the late 1970s, however, scientific evidence started to mount that inbred zoo animals had less chance of bearing young and the young they did bear had less chance of surviving and a greater likelihood of having some abnormality.

Then in 1977, at the Second World Conference on Breeding Endangered Species in Captivity, Nate Flesness, a researcher at the Minnesota Zoological Gardens, delivered a paper showing that the highly inbred Przewalski's horse—a wild Asian horse now completely extinct in its natural range in northern China and Mongolia—was statistically less successful at breeding than comparable but less inbred animal populations. The statistical reprimand irritated some in the zoo community. But the analysis stuck, and more followed.

Two researchers, John Ballou and Katherine Ralls, examined the breeding records of forty-four species of mammals at the National Zoo and they too found that inbred young had much higher rates of mortality.

As a result, when the AAZPA decided that conservation was to be its keystone, it realized that it had to conserve the genes as well as the hides of the animals. It was at just this juncture that Bob Beck wandered into the North Hollywood hotel. If he had come earlier, the story might have been very different.

"Captive breeding was initially tried in the late sixties, but the time wasn't ripe," said Thomas Foose, the AAZPA's former conservation director and now an official with the International Union for the Conservation of Nature. "Zoos weren't organized, data weren't available and the wildlife crisis wasn't acute yet."

By the time Beck sought the help of the zoo community, the "crisis" had hit with full force. Yes, islands, so delicate and susceptible to calamity, had suffered. But now entire continents were also threatened and the problem had worked its way into the scientific and popular consciousness.

"It became clear that humanity has been forcing species and populations to extinction at a rate greatly exceeding that of natural attrition," Paul and Anne Ehrlich wrote in their 1981 book *Extinction: The Causes and Consequences of the Disappearance of Species.* "In the last twenty-five years or so, the disparity between the rate of loss and the rate of replacement has become alarming; in the next twenty-five years, unless something is done, it promises to be catastrophic for humanity."

Of particular concern were the biologically rich tropical rain forests that belted the Earth. They cover only 7 percent of the Earth's surface, yet they contain roughly half of all the globe's plant and animal species. In the 1980s, rain forest was being felled at a rate of a 28 acres a minute, 40,000 acres a week, 23,000 square miles a year, an area about the size of West Virginia.

Some biologists calculated that the world might lose 25 percent of its species by 2015. Such estimates, however, were extremely soft, considering the fact that we have only a vague idea how many species are actually out there. Approximately 1.7 million have been identified, but biologists believe that there could be a total of anywhere from 10 million to 30 million. For example, on just one 2.5-acre patch of Peruvian rain forest, entomologist Terry Erwin found 41,000 different species of insects.

Most of these unknown species are obscure plants, bugs, and bacteria. What they all do and whether they are vital to humanity's well-being remain a mystery. All that is certain is that they are vanishing at rates unparalleled in the Earth's history.

Of course, along with the mites, the mighty are also threatened. By the next century, almost every animal that comes to mind when the word "wild" is used—lions, tigers, elephants, gorillas, bears—will be on the endangered list, if they aren't already. And that is what mobilized the zoo community.

Most species are threatened simply by the loss of a wild place to live. Some 67 percent of all endangered, vulnerable, or rare species of vertebrates are threatened by habitat degradation or loss, according to a study for the International Union for the Conservation of Nature.

But the threat doesn't come just from this wholesale destruction. The fragmentation of the wilderness into patches is enough to spell doom for many species. Patches where there aren't enough individuals to maintain a minimum viable population. Patches too far apart to allow for the exchange of individuals from neighboring populations. Patches that are a lot like islands, islands surrounded not by oceans but by humanity.

Not surprisingly, when trying to assess the risks posed by fragmenting forests, biologists turn to the work done on islands. Wilson, for example, used Terborgh and Diamond's work to calculate that a forest reduced to just ten square miles will lose between 50 percent of its birds during the next 100 years. The smaller the patches—Wilson warned—the greater the rates of extinction.

After habitat loss, competition and predation from introduced species of plants and animals and overexploitation are the two greatest extinction perils. Rhinoceroses are vanishing because of the demand for their horns for use as dagger handles and as an ingredient in love potions. Trees like mahogany and bois de prune blanc are at risk because of their market value.

The introduction of rabbits, gorse grass, or zebra mussels to a place where they simply do not belong turns a relatively benign entity into a monster. These alien species—referred to as "exotics"—are particularly insidious.

Unlike habitat destruction, where the simple act of refraining from felling forests or draining wetlands solves the problem, once an exotic plant or animal takes hold in a new habitat getting rid of it can be next to impossible. One biologist was moved to describe these aliens as "living pollution."

As a result of all these forces, it is estimated that the globe may be losing as many as forty-eight species a day—a rate twice as high as any time in the history of the world.

It was in the face of this monumental crisis that the AAZPA decided to create a "zoo ark" for at least a handful of endangered animals. Now, it appeared that passage on that vessel had been booked for the birds of Guam.

By 1992, there were forty-one endangered species of mammals, thirteen species of birds, six species of reptiles and amphibians, and one mollusk, a snail, being bred at 148 North American zoos and aquariums.

There were white, black, and great horned rhinos, cloud and snow leopards, Humboldt penguins, golden-lion tamarins, red pandas, white-cheeked gibbons, red wolves, Asian lions, chimpanzees, gorillas, and orangutans on the "zoo ark."

In 1981, the Bali mynah, a striking animal with white feathers, black wing tips, and blue cheek patches, was the first bird brought aboard the ark. Habitat destruction, development, and poaching are all eating away at the species. It is thought that perhaps only thirty to forty birds remain in the Bali Barat National Park on the island of Bali. There are, however, more than 350 birds in sixty-eight American zoos.

Every species on the ark has a species survival plan (SSP) and a "species coordinator"—usually a curator at one of the participating zoos. The coordinator is responsible for keeping track of the animals and for mixing and matching individuals from zoo to zoo, with an eye to preserving variety and genetic distinction.

The plan is overseen by "propagation groups" composed of representatives of the participating zoos. A stud-book keeper tracks the family lines and records the various moves of individuals from zoo to zoo for mating purposes. The entire exercise has been compared to computerized dating for gorillas.

Nearly 10,000 animals, in all, are on board the ark, from the huge African elephants to the tiny partula snail, which hails from the volcanic islands of the South Pacific. The snail lives in banana and bamboo trees and munches on decaying vegetation. It found its way into the species survival program because of the invasion of some exotic carnivorous relatives from North and South America. These larger, meaner snails prey on the little partula natives.

The zoos are committed to managing and housing some of these species, in theory at least, for as long as 500 years. As a result, each addition to the SSP list also means less chance for other animals to book passage, for the physical space is clearly limited. All the zoos in North America contain barely 20,000 acres and could comfortably fit into the New York City borough of Brooklyn.

Zoos, however, are zoos—institutions with institutional prejudices—and so in this competition for passage, a perusal of the SSP list reveals a strong preference for big, interesting animals—the partula snail excepted—that not only are endangered but also bring in the crowds as well. "Zoos are preadapted to specialize in the larger vertebrates," Foose concedes. But he goes on to argue that "those are the species that are going to be harder to sustain in the wild" and these animals "often serve as flagships for preserving other species and habitats. They are the charismatic emissaries of nature."

Yet as noble as the effort is, some critics question the cost and value. With literally thousands of obscure species facing extinction, they ask whether we are only deluding ourselves by worrying about a monkey here or a bird there. "Many of the less cuddly, less spectacular organisms that *Homo Sapiens* is wiping out are more important to the human future than most of the publicized endangered species," argues Paul Ehrlich. "People need plants and insects more than they need leopards and whales (which is not to denigrate the value of the latter two)."

All too often, valuable products—such as pharmaceuticals and foodstuffs—are the product of microbes, sponges, and roots. Similarly, the key to regulating something as vital as soil fertility falls to the myriad of creepy, crawling little things that break down cellulose and channel nutrients. It is unlikely that any of these organisms will ever have their own SSP.

Michael Soulé, a noted conservation biologist, has estimated that perhaps no more than 925 endangered species can be managed in zoos. Other analysts have been even less optimistic, placing the number at 500.

The tab for keeping a "minimum viable population" alive also runs into lots of seeds, geckos, and meat. One study placed it as low as $10,000 a year for Caribbean flamingos to $500,000 for Siberian tigers. Soulé has calculated that in the course of a 200-year voyage on the ark, a species could run up a $40 million to $100 million tab.

When considering such figures, some biologists wonder whether there aren't better ways to spend the money. For example, when Scott Derrickson, of the National Zoo, was asked to analyze the prospects for starting a captive breeding program for parrots—one of the emerging endangered families of birds—he concluded that "the best possible use" of the money might be in buying up rain forests in the Caribbean and Latin America. "Without that—habitat—where is a captive breeding program going anyway?" he asked.

In the case of the California condor, one of the most popular and most expensive of the zoo captive breeding programs, the habitat can't even be preserved. It has been consumed by oil derricks, farms, power lines, and suburban sprawl. The deserts, mountains, and forests the giant bird once considered home are gone forever. So, the SSP calls for limiting the range of the bird by modifying its behavior. The plan is to teach it to eat only carrion specially provided for it. The hope is that this will make the animal more sedentary. If successful it may preserve the condor by making it one of the largest pets in the world.

Another solution to the space and feed problem is the "frozen zoo." Here animal eggs, sperm, and embryos will be kept on ice and "mating" will be done in petri dishes and test tubes. Such "brave new zoos" already exist in San Diego, Cincinnati, and Washington, D.C. In February 1989, scientific history was made when a domestic cat gave birth to a male Indian desert cat in Cincinnati, the result of in vitro fertilization.

"We believe that the extinction of many species may be averted through the preservation of embryos and the accelerated propagation of animals," explained Betsy Dresser, director of the Cincinnati Zoo's Center for Reproduction of Endangered Wildlife—a state-of-the-art, $3.5 million facility built in 1991. But when a visit to the zoo takes on the same frosty aura as a stroll down the frozen food aisle, will the public be as enthusiastic?

And where does this all lead? "When you start choosing which animals live and which go extinct, you are starting to play God," Derrickson said. And what is really being saved? Frozen embryos and modified condors are hardly preserving nature. Furthermore, some species have been left trapped in captive breeding programs with even fewer prospects of release than the condor.

Some of these very issues would touch the Guam program all too soon. But on that February day back in 1983, Beck was not pondering the heated controversies or deep philosophical issues of humans becoming the stewards of nature. All he was looking for was help and he had found it. He returned to Guam brimming with optimism.

Chapter

5

April 1983

Julie Savidge had made an astounding discovery. Not one that was fit for the scientific journals, could speed her toward a doctorate, or solve the puzzle of Guam's disappearing birds. Still, it was a revelation and at the moment it was of prime importance—monitor lizards are crazy for peanut butter.

For almost two hours, a resolute monitor had been patrolling, probing, poking, and burrowing around the perimeter of Savidge's tent, driven on by the aroma of peanut butter. Armed with a tennis sneaker, the biologist had repulsed repeated attacks, but the monitor would not be denied.

Three to four feet in length and weighing roughly three pounds, *Varnus indicus* is Guam's largest lizard. Dark brown with lemon flecks, the monitor scuttles through the rain forests, climbs trees, feasts on carrion, and marauds chicken coops. It has even been known to take on a dog every now and then. Its bite is extremely dangerous, not because of any venom, but because of all the bacteria festering in the animal's mouth. One nip and you've got an infection that will defy many an antibiotic.

The monitor consumes birds and eggs and had been considered an early suspect in the decline of Guam's bird life. But Savidge had quickly dismissed the monitor. It and the native birds had obviously struck some natural balance over the millennia. Besides, a limited analysis of monitor stomach contents, done by DAWR biologists, had shown the animals had as much dining interest in spiders, shrews, snails, and crabs as in birds and their eggs. And now she had found one more thing the lizard relished—peanut butter.

Savidge had pitched her pup tent in the forests below the cliffs at Ritidian Point as a makeshift blind to observe one of Guam's last known rail nests. She had placed the tent about ninety feet from the nest and was peeking out of a small hole left in the zippered entrance. She had come to this spot many times in the past few weeks and patiently sat and watched for some clue as to what was happening to the island's rails. Would some predator come? Would there be some visible behavior to indicate that the birds were ill? Would something unforeseen show itself? For hours on end, Savidge would sit and watch and wait.

This time, while conducting her stakeout, she began to munch on a peanut butter and jelly sandwich. "I should have had more insight than to bring food into my tent," she said. "But this monitor smelled my sandwich while I was eating it and he was so insistent. . . . He started burrowing into the tent and I didn't know how to get him away. I'd take off my shoe and whack him on the nose. He'd retreat and then he'd come right back. I quickly ate my sandwich, but he could still smell it. We fought for two hours and then I finally gave up and left."

Monitors notwithstanding, things were going better for Savidge by that early March day. Bob Beck had returned from North Hollywood and was trying to build a captive breeding complex for rails, even though he readily conceded that he wasn't exactly sure what it ought to look like. The slides, nets, and stains had come and so had Julie's husband, Tom Seibert, who had landed a job at the University of Guam working on a project to combat an exotic weed that had taken root on the island.

The Chamorros called it *masigsig* and the botanists knew it as the Asian weed *Chromolaena odoranta*. The tough little plant, suspected of arriving in Guam as seeds in a ship's ballast water, had taken over fields and roadsides, smothering native ferns and grasses.

A similar infestation on the island of Trinidad had been arrested by a little yellow moth, *Pareuchaetes pseudoinsulata*, whose larvae ate the weed out of existence. "My job," Seibert explained, "was to get a colony of *Pareuchaetes* established on Guam."

While her husband was pondering moths and weeds, Savidge was slowly amassing the data, the hard evidence to back up what she suspected might be the story of another exotic species ravaging Guam.

Still, it seemed that everything the DAWR did was too plodding and whatever was happening out in the forest was hurtling toward the ultimate extinction of the birds.

Now there were hardly any birds left at Northwest Field or Andersen Air Force Base. Ritidian Point had truly become the last refuge. Clearly, there wasn't very much time left. Beck might be successful in rescuing species from the abyss, but if Savidge did not find the answer before the birds disappeared from the wild, she might never find the true solution. It would leave the annals of zoology with a mysterious massacre. It would also leave her without a Ph.D. thesis.

She had no choice. She would push on, keep to her plan, and hope that she could move more swiftly than whatever was out there in the forest. She would look for a disease and continue to hunt the snake. She would collect the data and let them fall where they might. Little did she realize into what a scientific maelstrom such an approach was about to plunge her.

She had spent hours drawing up a flow chart with her disease and predator hypotheses and the predictions that could be made from them. For example, if the snake was the problem, Savidge reasoned, the expansion of the reptile's range should coincide with the contraction of the habitats where birds were found. Could she find data that would show these two trends?

She hoped that her survey would help answer that question. Herman Muna was still driving around the island—from Potts Junction to Agat to Santa Rita—asking his twelve questions.

He would pull into a village, knock on doors and pose questions like:

When did you first notice a snake near your home? Have you seen rats near your home? Have snakes, rats, iguanas, cats, dogs, or pigs ever eaten your birds or eggs? Have your birds ever been sick? When did native birds start declining near your home?

In response, Muna gathered a spate of stories about the snake and remarkably many interviewees could virtually pinpoint the time they first remember seeing it. "It was the day of Carlos's baptism. I had gone into the garden to cut some mangoes and there was a snake on the tree." Carlos was now ten, so that was February 1972. It was that simple. Incredibly, guided by fiestas, holy days, weddings, and birthdays, most recollections in a village actually tallied.

But when it came to the questions about rats and disease, most people were nonplussed. Why were they asking about rats and illnesses? There was no problem. The snake was the problem.

"I got some 300 responses from all over the island," Savidge said. "What was interesting was the number of people who complained about snakes. Many of these people were local people, who raised chickens or pigeons, and a lot of them said they'd had to give up raising animals because the snakes ate them all."

She conceded that such a survey isn't really biology but defended the exercise, saying, "It was not a conventional situation; we had to use every tool we could think of."

But such unorthodox methods were already creating intellectual problems for Savidge, who found herself caught between the government wildlife officials, who paid her salary and judged her work, and the people of Guam. The officials were fixated on the presence of a disease, but the people who lived on the island insisted that the problem really lay elsewhere.

"You know," Savidge told her husband one evening, "the people, they really think snakes are *the* problem."

The DAWR and the U.S. Fish and Wildlife Service continued to suspect disease, so that remained Savidge's first line of inquiry.

She had found a bunch of bird carcasses in the freezer at the DAWR and sent them off to the Wildlife Health Research Center in Madison. Any other carcasses that she could get her hands on—from road kills, accidents, poultry pens, or shore bird colonies—she also sent along to Madison. The remains of a yellow bittern, a lesser golden plover, a Marianas fruit dove, a pintail duck, an egret, two frigate birds, and several chickens all made the trip to the Midwest on ice.

In December, Savidge had taken a ferry across a two-mile lagoon at the southern tip of Guam to a thin sliver of a coral atoll, called Cocos Island. The shores of Guam's rippling, green volcanic hills loomed over the little island. The waters were mottled in verdigris and indigo. But what was most distinctive about the scene was the black dots that swooped over the water and soared over the islands. Here the world was alive with the flutter of wings and the song of birds. Cocos was home to dozens of Micronesian starlings, doves, and terns.

Savidge netted sixteen black-feathered, yellow-eyed little starlings, each about eight and a half inches in length. These she took back to the DAWR lab and held in cages for thirty days. She took blood smears and then the birds were transferred to a large outdoor cage in a wooded area in central Guam. Half the cage was mosquito-proofed using an inner layer of mosquito netting and an outer layer of screen. The other half of the cage was open chicken wire.

This was a crude "sentinel" experiment to see if the exposed group would catch the disease that was plaguing Guam. Six starlings were protected and ten left exposed. Thirty days later, the birds were brought in. All were still in perfect health. Savidge then took them back to the lab and observed them for another sixty days. Still perfectly healthy. Then she took more blood smears and released them into a ravine forest on the island. Another attempt to find the Guam disease had been fruitless.

As the months ticked by and the dry season began to give way to the torrential downpours of the rainy season, Savidge scoured the woods for birds and blood samples. A few weeks after the battle with the monitor, she rose at 4:30 A.M. and drove the deserted roads from her apartment in the little crossroads village of Toto to the DAWR compound. There she met Muna, and together they set out in a DAWR truck to go catch some birds.

Their destination was a forest patch on Northwest Field, hemmed in by the old asphalt ribbons along which thirty-eight years before B52s had roared on their way to bomb Tokyo. Now the field was largely defunct, the forests dense and tangled, the

pavement cracked and weedy. Yet even in their present condition, the old runways were among the best roads on the island
and so Savidge and Muna sped along them, racing the sun to
their work.

The day before, Muna had hacked and hacked at the dense,
viny woods with a machete to clear space for the four nylon mist
nets they would use to trap the birds. Then he and Savidge had
set the thirty-foot aluminum poles and left the soft, filmy fabric
rolled up on them. Now, they were hustling to unfurl the nets,
thirty-six feet long and nine feet wide, before the sun's light and
warmth propelled the birds on their hunt for breakfast.

The sky was clear, but the forest was coated with morning
dew, and the work left both of them soaked from the knees
down. By six o'clock the four nets were set. Savidge would have
liked to have "run" more but that was all she had. She was still
waiting for a delivery of additional nets. Waiting patiently now,
on Guam time.

For the next four hours, Muna and Savidge patrolled the mile
or so of forest over which the nets had been rigged. They
caught no birds. Savidge pulled a notebook from her pack and
wrote: "Herman and I set up four mist nets at Northwest Field
without catching a single bird. I think the problem is too few
birds and not enough nets."

By 10 A.M. the sun was high and moving higher. The forest was
hot and steamy, and any birds would have found shelter and
hunkered down for a siesta. The next hour was spent rolling up
the nets, breaking down the poles and loading the gear into the
truck. The air was now so thick that it felt like hot, wet towels
had been draped over their arms and shoulders.

Shortly after 1 P.M. they were back at the DAWR offices,
empty-handed. "It was basically a seven-hour day, for which I got
nothing," Savidge said. Most troubling, however, was the increasing sense that the birds on Northwest Field and Andersen
AFB were dwindling. Savidge was worried they might already be
gone.

By early May, her order of nets had arrived and she and Muna
tried running ten of them at Pajon Basin in the Ritidian forest.
These had also been set the night before and now the two
moved quickly through the forest unfurling the snares and hop-

ing for the best. Four hours later, they had nabbed one female Micronesian kingfisher and one male broadbill.

When a bird was caught, the drill was precise and speedy. Savidge would disentangle the animal from the net, and give it the quick once-over for any signs of disease or physical trauma. Then she would plunk it in a soft cloth bag, flip open her tool box, pull out a little scale, and weigh the bag of bird. After that, she removed the bird from the sack and, using a little pincers, nicked the bird's toenail and smeared several glass slides with blood. Finally she would swab the oral and "fecal" areas with Q-Tips for bacterial and viral samples. The slides and swabs were deposited in a small, red Igloo cooler. If the toe was still bleeding, Savidge would daub it with a styptic pencil, like the ones valued by men who never quite master shaving. The bird was then freed. The entire exercise took no more than five minutes.

That same month, Savidge took her nets on a trip to Saipan, a Marianas island about 120 miles to the north. In a single day there, she caught sixteen bridled white-eyes with four nets. The birds were brought back to Guam in mosquito-proof cages and monitored in the lab for a month; then eight were placed in a wooded area on northern Guam for a second sentinel study. After a month, they had gained weight but had not gotten sick.

Savidge had dutifully made up blood smears of all the birds she had trapped and shipped them off to Madison. Her own perusal of the slides offered no evidence of disease and neither did the initial necropsies. None of the blood slides showed anything and the carcasses revealed nothing either.

Back in Madison, Louis Sileo, a senior wildlife pathologist, reviewed all of Savidge's samples. There were hundreds of blood smears, but Sileo, sitting at a big, black microscope, studied every single one.

"You sit there in front of the microscope and move the slide back and forth looking for red cells that have parasites in them," Sileo explained. "I found nothing—not a single parasite, which is very unusual. So unusual that I was worried that I was missing them. So, I took the whole batch of them to an expert on these matters, Ellis Greiner of the University of Florida at Gainesville. He said they were the most monotonous series of slides he'd ever looked at. He never found anything either."

•   •   •

The brown tree snake began increasingly to occupy Savidge's thoughts. There was still no hard evidence to prove that it was the culprit. But there was no evidence to prove it wasn't. One thing was clear, the nocturnal, tree-climbing reptile did eat birds and bird eggs. Still, could it eat enough birds to cause complete extinction? Could it eat all the birds?

Less than two miles down the road from DAWR headquarters lies the campus of the University of Guam. It is a set of utilitarian, block buildings with an emphasis on surviving a typhoon rather than making an architectural statement.

Housed in one of these concrete bunkers is the Micronesian Area Research Center, which boasts a set of khaki-colored steel file cabinets filled with a wealth of old press clippings about Guam and the other Micronesian islands, with some of the stories dating back to the 1940s. Savidge, now acting more the historian than biologist, combed the files for references to the brown tree snake.

It turned out that snakes were such an oddity on the island—Guam isn't supposed to have any save for a small, blind ground snake that could easily be mistaken for a worm—that any brush with the animal appeared to be worth a mention in the newspapers.

The earliest clipping was from a U.S. military post newspaper in September 1958. It read:

"There are no snakes on Guam." That is a statement heard often enough except at the U.S. Naval Magazine, where naval and civilian personnel have witnessed the capture of at least five reptiles on or near the station.

The most recent captive was 8.5 feet long, the largest to date. Identified as a colubrid (by Dr. O. W. Limig of the Guam Dept. of Agriculture).

They are not considered dangerous to humans because the teeth are set far to the rear of the mouth. But because of the size of the most recent specimen, it is believed that the snake could strike at a small child's lower leg and bring its grooved teeth into action.

• • •

The story was accompanied by a picture of two young soldiers—D. A. Collet and Witt Humbles—holding the snake.

Two months later, the *Guam Daily News* ran a picture of a nine-foot snake caught by two youngsters in the town of Agat, just south of the large navy complex, which sits on the southwestern shore of the island.

In early 1960, there was a picture in the newspaper of five Agat youths and the seven-foot snake they caught at a local gas station under the caption "Wiggling Prowler."

The caption noted, "Similar snakes have been found in various places and are believed to have been unwittingly 'imported' since the postwar period aboard ship."

In 1965, there were reports of snakes being caught at Santa Rita, a little town that is hemmed in by the Naval Station and the sprawling Naval Magazine, and in Apra Heights, just to the north of the magazine.

On June 24, 1966, Mrs. Edith Smith of Apra Heights awoke at 4 A.M. to discover a snake sliding across her neck, the *Guam Daily News* reported.

In 1967, one snake was caught at Sumay—right on the Naval Station—and another at Santa Rita. During the next four years, another six snakes made the newspaper, all found on the southwestern part of the island around the Naval Station and Apra Harbor.

By March 1971, the snakes had become a sufficient nuisance to the folks living in Apra and Apra Heights that a local legislator, Sen. James Butler, proposed a snake bounty. But the idea was opposed by local authorities. S. A. Perez, director of the Department of Agriculture, assured a Senate hearing that although the animal—which he called the Philippine rat snake—"where it is numerous . . . can be a visual nuisance . . . [it] does not pose a threat to man's physical health."

James B. Branch, a science consultant for the Department of Education, argued that the snake was a good way of keeping the island's rat population in check. "The Department of Agriculture, the science community, and the Department of Public Health do not consider Guam's snakes as dangerous pests, but do consider rodents as one of the island's major health hazards," he said.

Butler, faced with such astute opposition and general lack of interest on the part of his fellow legislators, dropped his campaign, although he pointed out that the snake feeds on coconut crabs, a delicacy much loved by the islanders, and the Guam rail, or koko. "The fact that the koko is the national bird of Guam ought to be enough to condemn the rat snake. If for no other reason than it's unpatriotic," Butler argued, tongue in cheek.

By 1975, snakes were being found in Asan and Tamuning— ten miles to the north of Agat, according to the clippings.

By themselves these stories didn't reveal very much. The snake was a curiosity and was almost always treated as such in the newspaper. But combined with the questionnaires and the DAWR index cards of field observations, they were pieces of a story about still another invasion of Guam. The Spanish, the Americans, the Japanese, and then the Americans again had all stormed Guam, but this was a more surreptitious, patient invasion.

Assuming that it did arrive as a stowaway in war materiel being shipped back to the United States, sometime after World War II, its point of entry would have been Apra Harbor in southern Guam.

Then the snake slowly moved from the harbor, Savidge calculated. First to the nearby Naval Station and Magazine—both thickly wooded, remote areas not easily accessible to the public. Here the snake established itself and then worked its way south and east the two to three miles to Santa Rita, Agat, and Apra Heights. By the 1960s, it had traveled seventeen miles across the southern part of the island to the eastern shore at Inarajan. By 1977 it had worked its way twelve miles north to Tumon Bay, and by 1982 it had arrived at the northern tip of the island, Ritidian Point—some twenty-five miles from the Naval Station. At least, that's what the newspaper clippings and DAWR biologists' field notes seemed to indicate.

Could the snake's steady progress across the island be linked to the decline of the birds? Savidge decided that she needed to learn more about the brown tree snake.

· · ·

Where to begin? Savidge was an ecologist, not a herpetologist. So, she began by simply looking for snakes. This was not as easy as one might think. The brown tree snake is nocturnal and cryptic, which is to say it is very good at hiding. So good, in fact, that even an experienced herpetologist staring at a thicket can miss an animal within arm's length. Still, she started by checking the logical place for a tree snake—the trees. To scout the crowns of palms, she borrowed a ladder truck with a cherry picker from the air force and sent Herman Muna up to look for snakes among the fronds.

"The bucket," Muna recalled, "was reasonably deep, maybe three feet, but to get into the crevices of the branches you'd almost have to get out of the bucket and there you'd be, fifty feet up."

"There was a certain Rube Goldberg aspect to what we were doing," Savidge allowed. The air force volunteer who was operating the rig must have been learning on the job. "We were swinging Herman around," she said.

Eventually, Savidge abandoned her tree-top search and reported in her field notes: "No good. Hard to manipulate. Dangerous."

The Hammond Barnhart *Dictionary of Science* defines "the scientific method" as "an orderly method used in scientific research, generally consisting of identifying a problem, gathering all pertinent data, formulating a hypothesis, performing experiments, interpreting data, and drawing a conclusion."

Field science, however, is not like the science performed on the laboratory bench. It is the place where method meets Rube Goldberg. Still field researchers consider themselves the equal of their laboratory-cloistered colleagues and observe the same rules, even if they don't play the same game.

Two of the most sacred criteria of the scientific method are "replicability" and "statistical significance." No scientific work can be regarded seriously without meeting both. Together they separate the facts from the wild claims.

Any scientist who claims that an experiment proves a hypothesis has to describe the materials and methods used and the results obtained. Then any other scientist ought to be able to replicate the experiment and get the same results.

The results themselves have to be based on a sufficiently large set of data and pass mathematical tests that confirm that the outcome didn't happen just by chance and is "statistically significant."

It is difficult enough to apply these standards in the laboratory, let alone in the field. But apply they do. "You aren't just wrestling with an intellectual problem, you are fighting the elements too," Savidge said.

Field biologists are often up early or out late—depending on the natural history of the animals they are studying. They are exposed to the vagaries of weather and luck. They may have to fight a large lizard over a peanut butter sandwich or may get lost trying to survey terrain. Entire experiments can be ruined by a tree falling, untimely torrential rains, or test animals accidentally escaping or dying.

While all this is going on, the field scientist is trying to gather enough clean, uniform data to meet the statistical tests. "I once had a discussion with a statistician who insisted that every experiment has to have several reproducible statistical sets. Well, how do you do that when you've got a single population of an endangered species and that's all there is in the entire world?" Savidge asked.

Indeed, it was hard to see how another scientist would be able to replicate what was happening on Guam. But the standards hold for field science, and whether it was glass slides stained with birds' blood or animal body weights or snakes plucked from palm trees, Savidge was determined to gather her data. "I think that people who do field research actually like the added challenge, the obstacles. You have to be inventive and you have to be tenacious," she said.

Muna was Savidge's intrepid comrade in these adventures in field biology. "I got hired through a federal work program," the young Chamorro explained. "When they told me I'd be working for the Department of Agriculture, I thought I'd be working with cattle."

Instead, Muna was posted to the DAWR and found himself climbing trees, swinging in cherry pickers, scaling cliffs, and trekking though the forest with pounds of equipment on his back. To Muna, a one-time runner-up in the Mr. Pacific body-building tournament, these were all tests of machismo and each one he gamely took up.

"Poor Herman. He was assigned to me at the beginning because I was so overwhelmed—and the things I made him do!" Savidge said.

Not long after the cherry-picker episode Savidge found snake skins in the tree branches along the limestone cliffs at Ritidian and reasoned that the crevices and holes in the stone face might be an ideal place for snakes to hide. She got another ladder, a flashlight, and a long wire hook and sent Herman to poke and probe the cliffs. It turned out that the cliff was a veritable reptilian high rise.

Muna would climb up as high as thirty feet and stick his wire hook into a hole and, as often as not, pull into the hot sun from its cool, sleepy den one angry snake. This he would drop into a burlap sack.

Eventually, Muna was able to dispense with the hook. "It happened by accident," he said. "Julie had told me to mark the holes I had collected snakes from. So I was using a spray-paint can. When I sprayed one hole, two snakes came out from different holes and I almost fell off the ladder. They came shooting right out. There was something about the fumes they did not like." After that, all he needed was a can of spray paint. He'd gas the hole and just grab the snakes as they went ballistic.

Savidge collected lots of snakes and lots of snake data. How long they were. How heavy they were. Number of males. Number of females. And she began to dissect the animals to see what was in their guts. There were lots of things in there, including birds and bird eggs.

Perhaps she could set a trap. Savidge concocted another plan. "I put out these dummy nests and had a camera with a light beam over the nest. When a predator would trip the light beam, it would take a picture. The idea was to get a feeling for who was taking the dummy nest eggs," she explained.

Of course, this required hauling nests, eggs, cameras, and sixty pounds of lead batteries into the forest. Muna carried it all in on his back. "We used some of the same transects [Engbring] had used before—up at Ritidian, on the cliff line," Muna said. It was quite a hike both into the woods and up to the cliffs, to an area where root and rock mingle to create treacherous footing. "We put out these dummy nests and I'd have to lug these big batteries in a backpack through the jungle. It is very rough terrain: the moss is really slippery and there are lots of sharp limestone rocks, very rugged. It was quite a ways in and one slight misstep and you'd end up all cut up. I think that was the most dangerous thing I've done, much more dangerous than collecting snakes on the cliff."

And the fruit of these labors? "We got some fine pictures," Savidge said. "Pictures of hermit crabs, rats, and monitors." But no pictures of snakes.

During the summer Louis Sileo came out to Guam from the National Wildlife Health Research Center. On his way out, the pathologist stopped at Midway Island to investigate what local wildlife officials there feared was a recurrence of avian pox, a disease that had decimated the island's albatrosses five years earlier.

Sileo determined, however, that the problem this time was that the birds were ingesting paint chips and suffering from lead poisoning.

Now he was prepared to turn his attention to solving Guam's mystery. After the lab and dissecting rooms, after days of poring over a microscope looking at slides smeared with the blood of sick and dead animals, it was great fun to be out and about in the world.

Sileo and Savidge drove all over Guam. They hiked the forests, climbed the cliffs, visited villages. "I liked Guam. I liked the humidity, the fragrances. I like working in outposts. I like working with field biologists," Sileo said. But while he was having a fine time, it seemed to him unlikely that there was a disease lurking on the island. "Intuitively, it didn't feel right to me," he said.

One afternoon, Savidge and Sileo drove down to the southern end of the island to get blood samples from black francolins bagged by hunters at a DAWR game station.

The "station" was nothing more than a wooden table and a couple of metal folding chairs along the side of a road with a big green plywood sign that proclaimed, STOP—GAME CHECK STATION.

Here hunters were required to record their spoils from Guam's limited season on the black francolin—a bird introduced to the island from India in 1961 for sport shooting. In one of the many conundrums of the island's bird life, the francolins seemed to be doing fine. This, of course, led to the theory that perhaps they might be the carriers of the disease wiping out the native birds.

While a DAWR biologist dutifully recorded the number and weight of the birds bagged by hunters, Savidge and Sileo clipped toenails for blood samples and with Q-Tips swabbed oral and anal areas for bacterial and viral samples.

Yet, Sileo couldn't help but feel that perhaps this was wasted effort. "Tell me, Julie," he said, "what do you really think is causing the birds to disappear?"

And even though they stood alone on an open road, flanked by broad grassy plains, with only the high, blue sky and the ever-present banks of white clouds as company, Savidge could only whisper her thoughts.

"I just couldn't bring myself to say it out loud," she said. "It seemed so far-fetched."

Savidge told Sileo, "I think it's the snake."

Chapter

6

October 1984

*E*very seat in the auditorium at New York City's American Museum of Natural History was filled, and still they came. It was late on a Friday afternoon—ten papers on conservation and endangered species had already been given—but people remained, lining the walls.

Since the Delacour Symposium eight months earlier, word of Guam's vanishing birds had been spreading through the scientific community, and now, at the centennial meeting of the American Ornithological Union, a paper entitled "Reasons for the Decline of Guam's Avifauna" was on the agenda. Between a talk on condor immunoglobins and a slide show on the endemic birds of the Dominican Republic, Julie Savidge took the podium to give voice to what she could only whisper to Louis Sileo a couple of months earlier.

As she stood there, Savidge looked out over a hall filled with some of the foremost people in the discipline. She outlined the problem—the steady reduction of avian populations. She described the efforts, so far fruitless, to find a virus, bacterium, or parasite that might be the cause of this decline.

Then she took the plunge. She showed a series of time-lapse maps plotting the distribution of birds and the distribution of the brown tree snake. The snake's expansion and the birds' decline moved like tango partners across the island. She noted that electric outages caused by snakes hitting power lines had increased eightfold in the five years between 1978 and 1982.

Then she discussed the gut-content analysis of the snakes, which showed that they consumed birds and bird eggs with great regularity. She conceded that the evidence, so far, was only circumstantial. Nevertheless, the data tended to reject disease—which had been the explanation of choice for federal officials and most scientists—and implicate the brown tree snake in this bizarre avian massacre. Nothing like this had ever been proposed before.

The *Wilson Bulletin*, the nation's foremost ornithological journal, said Savidge's presentation "took by surprise a standing-room-only audience. . . . Few could believe that a mere snake was that efficient a predator and could build up the numbers commensurate with such devastation."

"I remember the paper," said the Bronx Zoo's Brunning. "We were all very concerned about Guam, and when Julie Savidge proposed the snake . . . well, it didn't seem to make a lot of sense."

One of the most attentive listeners in the audience was H. Douglas Pratt, an expert on Pacific tropical birds and a skilled bird illustrator. Pratt was coauthor of *The Birds of Hawaii and the Tropical Pacific*, the most comprehensive guide to Pacific island birds. In fact, it was Pratt who published the first article noting the serious decline in Guam's birds, back in 1979.

Pratt had first gone to Guam in 1976 to work on his field guide. "We got to Guam two weeks after a major typhoon had gone through and really ruined the island," he said. "We were wondering if there'd be any birds. But birds seem to weather typhoons pretty well. Island birds do. And there were birds, in good numbers, in the northern third of Guam. But in the southern two-thirds, which had plenty of good habitat, and looked perfectly fine, there were just no birds there. I mean none.

"We went back two years later and went back to the same spots we'd visited in 1976. . . . We couldn't find any birds at all in a lot of places where we had found healthy communities two years before. Everything had disappeared."

At the northern end of the island there were still populations of all the species, but it was increasingly difficult to find them.

"It was like there was this advancing front moving up and anywhere on one side of it there weren't any birds and on the other side of it the birds were doing just fine," Pratt said. "It was a mystery to us. It was not just a decline it was a total disappearance."

And now Julie Savidge was standing before him with the proposed culprit in this zoological whodunit. Pratt, however, was far from satisfied.

"She gave her talk," Pratt said, shaking his head at the memory, "and I just sat there thinking this is all bull." But he did more than just think it. From the floor he challenged the very idea. "I've been on Guam several times and I've never even seen a brown tree snake," he snapped.

It wasn't just the visible lack of snakes that bothered Pratt. The theory was, he felt, deeply flawed and "inconsistent with the facts." In the first place, a snake had never been tied to a mass extirpation like the one on Guam. There wasn't even any evidence to show how a snake could do it. "For a single predator, and a snake at that, to wipe out an entire avifauna, that's hard to see," Pratt said.

Second, if the snake arrived in the late 1940s or early 1950s, why had it taken two decades for the problem to surface?

Third, according to Tino Augon's calculations, both the tiny white-eyes at the tree tops and the big rails down on the forest floor were being wiped out by the same predator. Could the snake be that efficient in the canopy and on the ground?

Furthermore, for the most part it seemed that the smallest birds were most vulnerable and "that didn't seem to make any sense from a predation standpoint," Pratt argued. "In general, predators are going to try to get the most they can for an expenditure of energy. So they are going to go for the largest thing they can catch."

Why then weren't the doves and kingfishers disappearing instead of the white-eyes and flycatchers? "That bothered me," he said.

Finally, not all the birds were disappearing. Many of the introduced birds could still be found in reasonably good populations, even when the native birds had vanished. "The Philippine turtle dove was doing fine. The Eurasian tree sparrow and the chestnut mankin were doing fine," Pratt said.

The Philippine turtle dove presented a particularly difficult stumbling block to the snake hypothesis. Savidge's fellow DAWR biologist Paul Conry had recently begun studying the dove and what he had found seemed to fly in the face of her theory.

Conry's surveys showed that Philippine turtle dove populations were increasing in the very same areas of northern Guam where the native birds were disappearing. "So you had this situation where one species of bird was going up and all the other species of birds were going down," Conry said.

Conry, in explaining his research, seemed almost embarrassed that his data disagreed with Savidge's hypothesis. But he said: "At the time I got to Guam the primary hypothesis was that the cause of the bird decline was avian disease. Avian disease is a problem on Hawaii. The native birds [there] are crashing and they've got introduced birds that are thriving—a very similar situation to Guam. It looked like the same."

Pratt had done his doctoral dissertation on a reclassification of Hawaiian birds and was well aware of the avian disease problem. "My first reaction was that there was some sort of bird disease on Guam, because it had all the looks of an epidemic, and I was primed for that belief by my experience in Hawaii because disease is an important limiting factor for birds there," he explained.

"Some of the areas we visited were places where the camps had been for people evacuated from Vietnam. The Vietnamese refugees had been on Guam temporarily before they moved on elsewhere and I figured, 'Well, something came in with them.' It just seemed like the timing was right."

Besides, there was something too simple, too pat for Pratt in the snake hypothesis. Islands, he had learned, are enigmatic places. Things happen that defy explanation.

"You go to an island like Rarotonga, the main island of the Cook group [about 3,000 miles south of Hawaii]," Pratt offered. "It has a fruit dove, a flycatcher related to the Guam flycatcher, and a starling. . . . Three breeding birds. But the flycatcher there has just declined and declined and declined. It is down to one little valley, one little stream area. There are fewer than fifty pairs. Up in the valleys it is essentially a pristine, untouched forest. There is nothing wrong with it. Why aren't the birds there?

There are these unseen, mysterious things going on islands."

And so Pratt poured forth his disbelief. "I remember being the vocal opposition," he said. "I figured I got on Julie's fecal roster." Pratt's criticism was so fierce that the editor of the *Wilson Bulletin* was moved to write, "Seemingly in a state of shock, some of the audience, during the question period, asked questions or volunteered statements that were unscientific, unchivalrous, and embarrassing to the rest of us."

At the podium, Savidge stood stunned. "Up until that point, I had never had anybody challenge me publicly in a big room and Pratt was kind of aggressive in his challenge. So, that kind of surprised me. I don't know if it shook me up, but it did catch me off guard.

"I felt, 'Good grief, there is a faction of people out there that I've got to convince.' But that just made me more determined to get as tight a story as I could, because I knew now there were people out there who weren't going to be convinced unless I did."

After the presentation Savidge approached her new nemesis. "What data can I get to convince you?" she asked. But Pratt's dander was up now. "No data," he replied. "There is nothing you could tell me that could make me think a snake could do this."

The centennial meeting of the American Ornithological Union ended with a gala banquet. Cocktails were served among the Museum of Natural History's dinosaurs, and the black-tie dinner was served in the whale hall. The ornithologists danced into the night.

Julie Savidge returned to Guam with the stakes significantly higher than she had expected. "I knew that I had to get as much data as I could to convince the skeptics. That worried me. It created more stress, definitely."

Soon afterward, in an assessment of Guam's endangered species proposal, Pratt was to write, "The 'snake hypothesis' is fraught with innumerable and insurmountable logical inconsistencies." When Savidge saw the critique she typed it out on a slip of paper and taped it over her desk. "Whenever Julie would get down, when things weren't going well, she'd look at Pratt's words and that would charge her batteries," Beck said.

Pratt, however, wasn't the only doubter. One morning back on Guam, Savidge received a phone call from Ricardo Bordallo, governor of the island.

Savidge's snake questionnaire had caught Bordallo's interest. "He told me not to worry about the snake," Savidge recalled. "The snake was no problem, but to look at pesticides." Military pesticides, Bordallo thought, were poisoning the island.

Bordallo followed the call with a memo to the head of the Department of Agriculture. "I would like these people to be advised to look into the chemical poisoning of this island. Let me repeat. Chemical poisoning, DDT and stuff like that, softens the eggshells of the birds, causing the embryos to die before they are mature," Bordallo wrote.

Even at the DAWR, Savidge had trouble convincing people. One afternoon, she and some of her fellow biologists were sitting around discussing the state of affairs on the island and the prospects for saving the endemic birds. "Somebody said, 'Gee we still don't even know what's causing the decline,'" Savidge remembers, "and I could feel this rush of adrenaline come over me and I said, 'Too bad you don't, I do!' and I walked off."

Chapter

7

January 1984

*H*ow do you stun a gecko while flying at 35,000 feet without arousing the interest or ire of the other passengers? This was the problem facing Bob Beck as he sat in a Pan American jumbo jet flying from Guam to Honolulu.

The great Guamanian bird lift had begun. Beck was shepherding six pairs of Micronesian kingfishers to the Philadelphia and Bronx zoos for the start of the long-awaited captive breeding program. There they all sat in the last row of the passenger cabin on an overnight flight across the Pacific. Pan Am had graciously "comped" the birds in the name of nature conservation.

The cabin was dark and the few passengers were mostly dozing. Time for the kingfishers' late-night gecko snack. "We had no choice but to feed the kingfishers on the transoceanic flights," Beck explained. "They have such a high metabolism and such a low body weight that they just had to be fed" and the geckos proved to be an ideal in-flight meal. "They were a wonderful packet of food for these guys, containing liquid and everything," he said.

Beck and Muna had filled a large jar with 500 geckos for the trip, but before the little lizards could be dropped into the small plywood and wire-mesh cages now strapped into passenger seats, they had to be rendered senseless, so as not to create little tempests in each cage as birds chased skittering geckos—

possibly hurting themselves or rousing the sleeping passengers. But how to stun the geckos?

Jar in hand, Beck went into the lavatory at the rear of the plane. He closed the door, slid the latch, and unscrewed the jar top, cradling the glass in one hand. He fished out a gecko with the other hand and, holding it by its tail, swung it in a small arc than ended when the lizard's head hit the metal sink. Whap. That worked just fine. Whap. Whap. Whap. Beck emerged from the toilet with a dozen geckos swinging by their tails. Quickly, he dropped them into the cages. But a kingfisher, once it gets a gecko by the tail, it is naturally inclined to give it a few whaps itself. This they did until a little chorus of whap-whaps rose from the back of the plane. "What's that?" a fellow passenger asked Beck. "Oh, just some tropical birds being transported to zoos," he said. "No, the tapping?" he asked. "Just the birds," Beck replied.

So far so good. But then the calls started. It began in one box. "The kingfishers had this propensity: about every half-hour or so one male would sing, and the other birds would start responding," he said.

A male would sing out. "They had a real loud call—*breee, breee*—and they'd be singing like crazy and pretty soon the birds in the other boxes began answering and you'd get this chorus of kingfishers," Beck said.

"Most passengers would look around and were a little startled," he said. "But at one point this guy came storming to the back of the airplane muttering: 'I can't get any sleep. This plane is too damn noisy.'

"I thought, 'Oh, Christ, this is not going to be good.' No sooner had the guy lain down than the birds started calling. He jumps up—'I paid good money for this? What is this?' and stomped back to the front of the plane."

It was late the following evening when Beck's avian road show touched down in San Francisco. They would not leave for the East Coast until the next day. The Pan Am ground crew was ready to deposit the birds in the airport cargo storage area. "Well, you know how cold it can get at night in San Francisco, so I said: 'No way can we do this. They've got to be in a warm room.' But they didn't have any warm rooms," Beck recounted.

Someone suggested they use the cargo manager's office. The birds, to Beck's surprise and delight, were wheeled into a plush office. It was certainly warm.

The next morning at seven, Beck was sitting on the couch in this paneled and carpeted office, whacking geckos on the edge of the coffee table, when the cargo manager walked in. What happened next happened fast, but as best as Beck can remember it went something like this:

"The guy whose office this was, who knew nothing about this, walked in, took one look and freaked out. 'What in hell's name is going on in my office? I am a vice president of Pan American. I've got fucking birds in my office and some guy breaking lizards' heads on my coffee table.' And he went storming out of the office."

"Of course, it was my fault," Beck concedes, chuckling at the memory. "I should never have done that in his office. Killing geckos. Whack. Whack. Whack. Anyway five minutes later, after talking to people, he came back into the office and shook my hand. 'Very glad to know you,' he says. 'This is a wonderful project. Sorry I freaked out, but you've got to understand.'"

At 2 A.M. the next morning, more than forty-five hours after leaving Guam, Beck and the kingfishers finally landed at Kennedy Airport in New York City. Larry Shelton of the Philadelphia Zoo and Eric Adler of the Bronx Zoo met the birds. Adler took one pair to the Bronx. The ten other birds made the trip down the New Jersey Turnpike to Philadelphia. By the time Shelton and Beck had gotten the birds into their new cages at the Philadelphia Zoo's Penrose Laboratory, it was dawn.

As soon as Shelton had left Guam in September 1983, Beck went to the forests and began trapping white-eyes, fantails, and broadbills. He met with what appeared to be quick success, snagging several fantails and broadbills.

Both species are highly territorial, so simply playing a tape of calls was enough to propel the birds into the air to defend their turf. All Beck had to do was place a mist net across a break in the forest and wait. "When they think there is an intruder, they will fly around and around and eventually they fly into the net. It is a very simple procedure," he said.

The white-eyes were another story. While the fantails and broadbills lived in the lower part of the forest, the tiny white-eyes, who got their name from the white ring around each eye, inhabited the very top layers of the canopy. Just getting nets that high—more than forty feet off the ground—was a challenge.

After experiments with poles and tree climbing, the DAWR team found the most efficient method was to get to the treetops by hurling into the branches rocks with ropes tied around them, and then hoisting the nets on a pulley system.

But simply getting a net up there didn't assure capturing white-eyes, which are highly social and devoid of any territorial instinct. "What we tried to do was find where these birds were moving every day and then we'd put the net up in that location and hope they would come through randomly and intercept it. They didn't," Beck said.

He tried playing white-eye flocking calls but that offered no help. Beck was never able to catch a single white-eye. "It was a complete failure. The problem was that we were down to too few birds at that point," he said. Within a couple of months there would be none at all.

Soon Beck realized that it wasn't only the white-eyes that were lost. All the broadbills and fantails captured were males; the females had vanished. Fantails and broadbills still lived, but for all intent and purpose the species were dead.

Guam had once had eleven native varieties of forest birds. Among those were five species or subspecies found nowhere else on Earth—the Micronesian broadbill, the bridled white-eye, the Micronesian kingfisher, the rufous fantail, and the Guam rail.

Now three of those birds were extinct. Gone forever. The DAWR had sought to place these birds on the federal endangered species list back in 1978. More than five years later the application was still pending in bureaucratic limbo. But it would no longer matter for the broadbill, the fantail, and the white-eye.

The other species, such as the cardinal honeyeater and the

Marianas fruit dove, had also just about disappeared from Guam, although they could still be found on other nearby islands.

Shelton had left Guam just a few months before with visions of trapping, transporting, and captive breeding all the island's key species. It would have been a unique conservation project. But that dream was now crushed and Beck's worst nightmare was slowly coming true. He had tried to save the broadbills, fantails, and white-eyes. He had failed. "It was extremely disappointing and we were all really depressed because that was three species down the tubes," he said.

"For a little while we toyed with just capturing males from both the subspecies of the fantail and the broadbill. This fantail is closely related to fantails nearby. The broadbill also has closely related species nearby and we toyed with capturing the males and breeding them to these other species and crossing them back to the males, eventually restoring the majority of the gene pool.

"This was tried with the dusky seaside sparrow. We were actually going to try that but then we lost them all. All of a sudden they crashed," he said.

Whether even the cross-breeding could have save the bird is debatable, for the last true dusky seaside sparrow—a male—died at Disney World's Discovery Island, in Florida, in June 1987. Two years later, the last hybrid sparrows—the offspring of male dusky seaside sparrows and female Scott's seaside sparrows—vanished in a thunderstorm and the species was declared extinct.

"My feeling was if we had started six months sooner we would have had them, but we just started too late. If Julie and I had been hired six months sooner we would have had them," Beck said.

So, late in the fall of 1983, Beck and the DAWR focused their attention on the two remaining birds special to Guam—the Micronesian kingfisher and the Guam rail. "We turned to those species with a vengeance because we didn't want them to go down the tubes. We also knew that we had a lot more kingfishers and rails," Beck said.

Now, all the tramping through the woods that Augon and Beck had done for the past year paid dividends, for to trap these birds the first thing you had to know was where to find them.

"The kingfishers weren't limited to one site below Ritidian Point. These guys were spread out over Northwest Field, the conventional weapons area [on Andersen Air Force Base], and the Ritidian and Pajon basins. It was a much more widely distributed area," Beck said.

"The real crux of the matter was to find the birds first and then delineate their territory and find their nests. The kingfishers required quite a bit of field observation and field work, plus we wanted to sample the birds over as wide a geographic area as possible so they wouldn't be so closely related, and we had a goal of capturing thirty to fifty birds, which we figured was a decent enough sample to preserve genetic variation and there were still enough birds out there to do that."

To find the kingfishers, Beck and Augon cut transects through the woods, stopping every few hundred meters to play tapes of the kingfisher's *bree, bree* call. Once a kingfisher was located, it was marked on the map and a careful examination of the surrounding area was made. "If they were there consistently, we'd go back and try to trap them," Beck said. "Generally we could pretty quickly find their nest cavities. Sometimes we'd just be in the forest so much we'd hear them.

"Some of their behavior, for example their building a nest cavity, is very vocal behavior. . . . They seem to make the nest cavity not only for breeding but for pair bonding as well. They make a lot more cavities than they use. Sometimes, you find a whole row of cavities going up a tree, most of which are not completed.

"Playing tapes and spending time in the territories until you figure out what's going on. It takes a lot of observation, a lot of time," Beck said. "After a while you learn what kinds of trees they favor. Far and away, their favorite is a species of tree called pisonia. It is a very soft wood. It gets very big and gets way above the canopy. But when it gets above the canopy, because of its soft wood, the typhoons break off its limbs. So the tree will be alive, the base tree will be alive, but its limbs will be dead. Very dead. The kingfisher likes to nest in the extremely rotten branches of this big tree."

The kingfisher needed these extremely soft limbs because

even though it had a thick, long, black bill that emerged from its cinnamon-colored head and hung out over its two-ounce body, it was not as hard and steely as the beak wielded by the red-bellied woodpecker or white-breasted nuthatch. Nor did it have feet that could grasp a tree trunk so that it could bang away at a limb to create a nesting cavity, the way a woodpecker can.

Instead, the male and the female find a perch near their chosen tree and take turns flying at the tree, giving a call and taking a couple of pokes at it with their bills. "They stand off maybe five or six feet from the tree together and start calling and then dive into the tree, take a chunk out, and fall back, and you can hear that from a great distance," Beck said.

Given such handicaps, it is clear that it would take an awfully long time to excavate anything but the softest wood. Still, nesting cavities had been found on Guam in banyan trees, coconut palms, and telephone poles, but the tree of preference was the pisonia, or the *umumu* to the Chamorros.

They were easy enough for Beck to find, for they send barren fingers of wood sticking up over the lush canopy. Below they have thick clusters of glossy, green leaves growing out of the stumps and trunks.

"One huge pisonia tree was such a super site, down at Ritidian, that every time we'd capture a pair off this tree another pair would move in from adjacent territories. So it was a prime location. I think we captured three pairs out of that one tree. We'd come back one day after another and pairs would move into the site in just one day."

The actual trapping was an exercise worthy of the Army War College. It began with setting up two sets of aluminum poles about forty feet tall at opposite ends of a break in the forest, a break often tailored by Muna with a machete and tree snips.

It took a four-man crew to raise and tether the nets and as many as fourteen nets were set at a time. Then a hunting party—usually Beck, Muna, and Augon—would march from net to net using a tape recording of *bree, bree* as the bait.

Beck might be positioned on one side of the net with a lunchbox size Sony portable tape recorder and a walkie-talkie; Augon would be on the facing side, also with tape recorder and radio;

Muna would be in the middle, near the net, watching, also with a radio in hand.

If a kingfisher flew by and missed the net going left, drawn on by Beck's taped bird call, Muna would radio Augon and tell him to start his tape recorder. He would also tell Beck to stop his machine. If all went well, the second tape would lure the bird back and this time it would get snagged in the net.

"We'd be hiding on opposite ends and a lot of times it was a breeze," Muna said. "We'd just catch the pair in that territory and we'd go on to the next, they'd go right into it. But sometimes they'd get smart on you."

"It isn't as simple as it sounds," Beck cautioned. "Some birds we were never able to capture. We knew their territories, but for some reason they would never come into the net." As long as there were birds known to be in a territory they would trap four or five consecutive days and even return and try after that. But in some cases the birds simply could not be lured into the net.

"Other times," Beck said, "we actually had several birds on a territory on one spot. We actually were able to capture five birds in one day out of one net. The birds were drawn in from adjacent territories once we had captured the pair off the immediate territory. That was a pretty astounding day."

These trapping days, like Savidge's, began at 4 A.M. at the DAWR and ended about noon. Then the crew would go back to the divison complex and feed and care for the birds already in captivity. After that they would spend their evenings at a local shopping center's parking lot hunting for skinks and geckos. At ten o'clock at night, there would be two or three grown men, running around the asphalt and cruising the fences, chasing lizards to feed the birds the following day. (The little lizards simply refused to be bred in captivity.) Up late catching geckos, up early trapping birds. "For a while," Beck said, "it was a real grind."

By January six pairs of kingfishers had been caught, and Beck decided to head east. "We decided not to wait until we had the thirty birds. We'd do it in increments," he said. The decision was prompted, at least in part, by the difficulty of maintaining the birds in captivity. "When we brought those six kingfishers to the Philadelphia Zoo that was fifty skinks or geckos a day. That became a very oppressive task. We tried artificial food. We tried mealworms. But nothing satisfied them like skinks and geckos."

It was also decided that the birds would fly in the passenger cabin and Beck would accompany each shipment. "We didn't want to run any risks with these birds. They were just too rare. We didn't know how many more we could capture or how hard a time we'd have with them on a flight, what problems we'd run up against, plus we didn't know what would happen when they got to the Philadelphia Zoo. Were these birds going to be super-sensitive to diseases in the States? Were they going to have no immunity at all? So we were just really cautious about everything. It was a really scary time," Beck said.

As difficult as it was to catch kingfishers, bagging the rails turned out to be an even more challenging exercise. The goal was the same as with the kingfishers, to capture thirty or more genetically distinct animals. But catching rails was going to be a completely different exercise because they were speedy but flightless forest ground dwellers. The Chamorro had hunted them for food, but no one had ever thought much about trapping them alive.

"Nobody really knew exactly how to capture rails," Beck said. "Local people told us you could capture rails in traps with bait. But we never had any success with that."

Muna had grown up in the village of Gigo, not far from Andersen AFB. "Rails were everywhere. I grew up eating rail," he recounted. "Back then it was a big competition, who had the best rail dog. My father's dog would take me out collecting rails. We'd just wait in one spot and the dog would just keep coming back with rails. The really good ones didn't even kill the rails, just hold them firm in their mouth. We could fill up an old fifty-pound feed sack full of rails. He was just an old booney dog, but I sure wish we'd had that dog when we were collecting rails."

Trapping the rails took on an added urgency when it became clear that they, too, were rapidly dwindling. One of the first things Beck had done after being hired was a rail survey, and he had found that unlike the kingfishers, the rails' range was already limited to sections of northeastern and northwestern Guam, primarily on military land. A year later, they were down to just a single area—the dense forested areas on Andersen Air Force Base. They had completely disappeared from neighboring Northwest Field.

After some preliminary efforts with calls and traps failed, Beck hit on the idea of a "rail drive" to simply flush them out of the woods. "We eventually ended up by spreading out large nets. We'd find a habitat patch that had rails in it. We'd cut a swatch, place down the net, weighing it down with rocks, and drive the rails into the net. . . . We asked the air force for volunteers and they gave us their rapid response teams, which are trained to scour crash sites for debris after accidents. . . . We captured some of our rails that way, chasing them into the net. Other rails we captured because during the rail drive we'd come across a nest with eggs and we'd take those and incubate them and hatch the rails."

Again this was no small production, it took thirty to forty beaters to form a cordon that would march through the tangled, vine-choked woods pushing the rails before them. "People would walk through slowly, pushing brush aside, checking for nests," Muna said. "It was hot and tiring work and a lot of people got bee stings. Julie got stung pretty badly doing that."

"You had to be fast," said Eugene Morton, a biologist with the National Zoo; "as soon as they hit the net you had to grab them. One rail busted right through the mist net and kept going."

It was hard work that required a big cast and the returns, at least at the outset, were very modest. It was really hard to catch rails.

Morton, an expert on bird calls, had flown out to Guam in February 1984 to try to help Beck find new trapping methods. "We had not been having much success with the driving technique at that point and Gene was convinced that we could go into the forest and he could play tapes of rail calls and the rails would come right up and we could capture them pretty easily," Beck said. "One of the things we proved that week is that you couldn't do it with tapes."

Morton did, however, capture some rails. One day while walking through the woods, Beck, Morton, and Muna scattered a little covey of birds. Each man went diving into the underbrush after one of the fuzzy little chicks. Each came out with a prize. In one serendipitous burst, they had caught nearly one-sixth of all the rails that would ever be brought into captivity.

And that might have been all they ever got, if it hadn't been for some old-fashioned political lobbying and logrolling back in Washington, D.C.

The story of the Guam rails and Washington power brokers began one day when Beck and Muna were up at Andersen reconnoitering the last rail habitat—a patch of woods next to a big asphalt square called "Nancy-Ram," where a bunch of B52s were parked.

Just two months before, a car loaded with dynamite had blown up a U.S. Marine barracks in Lebanon, killing more than 200 people. Now, American military bases all over the world were beefing up security measures against terrorists. On Guam, security had always been pretty relaxed. There weren't even fences around many of the facilities, since until 1962 no one could even get on the island without military clearance. But now the rail's patch of woods was deemed a risk to the B52s.

"The base commander decided he had to cut down the habitat because it obscured the fence from the bombers, so what you could do is hop the fence, hide in the vegetation, and sabotage the bombers," Beck explained. "We got word on a Monday morning . . . that they were going to clear this area, immediately, later that afternoon.

"We met with the civil engineering commander and his staff and talked about the problem and they said: 'We are under orders to clear this area. We've got to clear it immediately.'

"We pointed out that this was a proposed endangered species and because of that you had to do an environmental impact statement or assessment. They said they didn't care about that. They had to do it because it was a national security issue, which is what they always do."

The American military had paid a high price for Guam and Micronesia and as a result it naturally felt it ought to be able to do as it pleased.

The United States had lost the island to the Japanese almost as fast as Captain Glass had won it from the Spanish. On December 10, 1941, three days after Pearl Harbor, Japanese troops stormed ashore at 4 A.M. following two days of aerial bombing. Two hours later, Captain George J. McMillan, governor of Guam, surrendered.

The Americans reclaimed the island two and a half years later, but winning it back wasn't easy. For weeks Guam was pounded by naval artillery and aerial bombardment to destroy Japanese air power and soften up the 20,000 troops defending the island. Agana, the island's major town, a community of broad, tree-lined streets and buildings with peaked, thatched roofs, was completely obliterated

On July 21, 1941, 20,000 Marines hit Asan Beach, on the western side of the island. For five days a pitched battle raged, claiming the lives of more than 4,200 American and Japanese soldiers. At the same time, an equally large landing force was wading through chest-high water toward the beach at Agat, which was not far from the village of Santa Rita.

The landing sites were just north and south of Apra Harbor and the Orote Point airstrips. These were the prizes that the American military sought.

On August 10, the USS *Indianapolis* sailed into Apra Harbor to accept the Japanese surrender, just as Captain Glass and the *Charleston* had done back in 1898. But the toll this time had been much steeper. About 55,000 American soldiers had taken part in the battle. Nearly 1,300 were killed and 5,600 were wounded. More than 9,000 Japanese soldiers were dead and another 10,000 were prisoners. The American military had won back Guam. It was, the generals and admirals felt, theirs.

After the war, the Pentagon argued for the outright annexation of Micronesia, but instead most of the islands were placed in a United Nations trust, overseen by the United States. Guam, however, once again became a militarily administered territory. In 1945, Admiral Chester Nimitz said the island was "a top strategic project" and called for "the reconversion of Guam into a Pacific base second only to Pearl Harbor."

The air force and the navy appropriated about a third of all the land on Guam for military installations. The island's Land Owners' Association has complained bitterly about the seizures, which include some of Guam's best farmland, fishing areas, and drinking water supplies.

In 1950, administration of the island was passed from the Department of Defense to the Department of Interior, and thus began a long march toward self-government and possibly some day a form of independence. But for now the military still swings a lot of weight.

Of course, many other Micronesian islands have fared much worse. Several atolls of the Marshall Islands were vaporized in nuclear bomb tests and radioactive clouds drifting over the Pacific accidentally settled on villages and water supplies on nearby islands. Some Marshallese were forced to move off their islands so that the air force could shoot missiles from California and have them fall in the lagoons of Kwajalein Atoll.

As the cold war grew so did the military presence on Guam. It became the main Strategic Air Command base in the Pacific. The air force's 43rd Strategic Wing, twenty B52s with nuclear-tipped missiles, was billeted at Andersen AFB. The Navy also deployed nuclear submarines with Polaris missiles and P-3C AWS aircraft with nuclear depth charges on Guam. The 250 earthen bunkers at the Naval Magazine—one of the first places the brown tree snake made its beachhead—has been the United States' main nuclear weapons storage facility in the western Pacific, with an estimated 200 nuclear devices stored there in the 1980s.

As far as Pentagon planners were concerned Guam had a big role to play in America's geopolitical-military strategy. It was a key link in assuring shipping and economic access to the growing markets of the Asian rim and in containing Soviet expansion. It was, in fact, one of the most heavily militarized islands in the world.

This was reasonable. The military had paid for Guam in American blood. It was theirs. Rails or no rails.

"We pointed out to them that we wanted to capture the rails anyway," Beck said. "It is not as if we wanted to leave the rails there forever." All the DAWR needed was time to catch the birds; then the air force could fell every last tree and bush in the name of national security.

"They asked how much time we wanted, and we said that it was hard to say. This is not an engineering problem. It is a biological problem," Beck said. "We might get them out as soon as three weeks, it might take six months. We told them we would do it as expeditiously as we could. The guy said, 'I'll give you two weeks,' and we said that was not satisfactory. So the meeting broke up at that point. I remember a lieutenant colonel calling the birds, 'those damn, cretinous rails.' It wasn't a very friendly exchange."

They returned to the DAWR office and sat around bemoaning the "damn, cretinous" air force. Gene Morton listened to the lamentations. Perhaps he couldn't call rails, but maybe there was something he could do in Washington.

Two days later, Morton and two pairs of rails left for the National Zoo. A few days after he returned to the mainland, he went to the American Museum of Natural History in New York City for a meeting of the U.S. section of the International Council for the Preservation of Birds. There Morton gave a brief report on the status of the rails and the air force's plan to level the last remaining rail habitat.

The Bronx Zoo's Don Brunning, one of the original curators involved in the Guam project, was flabbergasted. He contacted the public relations officers at the zoo and had them issue a press release. Thomas Lovejoy, a vice president of the World Wildlife Fund, was also surprised and within days had mobilized his organization. Michael Bean, a lawyer with the Environmental Defense Fund, immediately wrote to the Departments of Defense and Interior warning that he would sue for violations of the federal Endangered Species Act if the rails were lost.

A few days later, the World Wildlife Fund, a key international environmental organization, held a press conference, in Washington, D.C., where Lovejoy tried to strike a reasonable tone, conceding that security was important but arguing that the rail was a case where "an exception could be made to what are generally valid concerns."

Lovejoy's comments and Bean's letters appeared in a front-page story in the *New York Times* the next day under the headline

"A Tiny Flightless Bird Stalls U.S. Strategic Air Command."

Bean's threat was still a bit premature, because even though nearly six years had passed, the Fish and Wildlife Service still hadn't placed Guam's birds on the federal endangered species list.

"We contacted every congressman and every senator we thought would be sympathetic. We contacted every member of our board who knew someone or knew someone who knew someone," Brunning said.

The combined weight of the press conferences, threatened lawsuits, and lobbying had the desired effect. For the moment, the rails were safe, at least from the air force.

Chapter

8

April 1984

On the afternoon of April 17, Julie Savidge and Herman Muna loaded seven odd-looking contraptions and seven quails into a truck and headed for a patch of forest at the edge of Andersen Air Force Base, near the village of Dededo.

The devices were cylinders of heavy plastic mesh about three feet long and a foot in diameter. At the mouths of the tubes were mesh funnels tapering toward the center. The hole at the small end of the funnel, just a few inches in diameter, was covered by a trap door that swung in but not out. The top was covered with a sheet of plastic. Long wires were attached to each end.

This invention was the product of a year's on-again, off-again work. But since returning from the stormy American Ornithological Union (AOU) meeting in New York such a tool had become crucial for Savidge and now she was filled with hope. If all went well, it would deliver the proof she needed to show that the brown tree snake was responsible for the disappearance of the island's birds.

The last five months had been difficult ones. She had returned to Guam the previous October to face the relentless disappearance of the birds and lingering disbelief from the scientific naysayers. "The 'snake hypothesis' is fraught with innumerable and insurmountable logical inconsistencies." Douglas Pratt's words hung there before Savidge's eyes every day.

"Proving" that the snake was the perpetrator, however, depended not on building a case on circumstantial evidence, but on actively testing the three basic hypotheses Savidge had formulated. Three hypotheses that, if the snake was truly the culprit, she should be able to prove.

The first of these was that the expansion of the snake's range on the island should coincide with the disappearance of the birds. This she was well on her way to establishing with her surveys, newspaper clips, and DAWR records.

She had calculated that since the first snake was recorded in Santa Rita in the early 1950s, the species had moved into southern and then central Guam. It then headed north in the 1970s and steadily advanced at a little less than one mile a year until it reached Ritidian sometime in the early 1980s.

The DAWR roadside bird counts and reports reflected a similar pattern in the disappearance of the birds—with the smallest species vanishing first.

In 1963, for example, a DAWR study indicated that white-eyes, flycatchers, fantails, and honeyeaters were already absent from the Naval Magazine and snakes had been sighted within two miles of the naval installation.

A second proposition was that since the snake in its native habitat also ate rats, shrews, and lizards, these populations should also be in decline in Guam's forests.

But the crucial hypothesis was that areas with birds should suffer less predation from snakes than areas where birds had been wiped out and snakes were established.

The problem, Savidge said, was that "the only way I could figure out how to do that was to have some sort of a trap. You don't know how many hours I spent trying to develop a trap that would work. Tom would ask, 'What's the trap of the week?' We tried everything. I struggled for eons and I'd get pretty disillusioned thinking I'd never get this to work."

Building a snake trap, like the proverbial "better mousetrap," was no simple feat. Savidge had begun naively a year earlier with a fishhook and a slice of liver.

She had put out a big fishhook baited with a piece of liver in a grove of ironwood trees where the branches were hung with garlands of dried snake skins—a sure sign that the animal was about. She came back the next day. "Liver still on hook," she wrote in her field notes. Perhaps the snakes didn't like liver—after all, who does? She tried other meats. Still no takers.

Soon, Savidge decided that her trap needed to use birds for bait because that would be the best way to simulate a population of birds being preyed upon by the snake.

She tried a live quail, its leg tethered to the hook. The snake, or some other animal, took the bait twice, killing the bird and regurgitating the hook. She tried a three-pronged hook. The bird was eaten and the hook regurgitated again.

Then came a five-gallon paint can with a screen trapdoor across the opening and a sparrow inside. The next day the sparrow was gone and the can was occupied by a well-fed monitor.

The paint can was followed by a modified rat trap. A long "tunnel" led to the standard steel spring trap. The next day, "the bird was missing, the rat trap had gone off, but we hadn't caught anything," Savidge said. "That was my problem. They were getting my bait, but I wasn't catching the little rascals."

That was as far as her tinkering had gone before she went to the AOU meeting in New York. When she returned, the stakes had been upped significantly, and the snake had become not only the focus of her work but an obsession. It was also somewhere around this time that Savidge got her first snake bite. Two brown tree snakes lived at the DAWR lab—Claudia and Big Guy—and it was Claudia who drew first blood.

There was, Savidge had learned, no domesticating a brown tree snake. They are ornery by nature and ready to strike at any opportunity. And as brown tree snakes went, Claudia was extremely foul tempered. "She was probably one of the meanest snakes. . . . She was real, real mean," Savidge said.

All this Savidge knew. But one day, she doesn't remember exactly when, while handling Claudia, Savidge was distracted, and before the young biologist knew it, the animal had sunk her needle-sharp teeth into the fleshy crescent of skin between the thumb and the index finger. Worse, Claudia wouldn't let go. Savidge needed one of her colleagues to peel back the snake's jaws in order to free her hand. The bite hurt and bled a little, but at least the snake's venom wasn't poisonous to humans. Savidge started wearing gloves when handling snakes.

Savidge wasn't the only DAWR staffer to have a run-in with

Claudia. In an effort to shift the snake's eating habits from night to day, Savidge began to keep Claudia in the darkened, windowless bathroom at the DAWR office. One morning, the division's public relations officer went in to the toilet to find that Claudia had escaped. The serpent was bobbing, weaving, and blocking the door. "We heard this blood-curdling scream. We raced in and there were Judy and Claudia," Savidge said.

In November 1983, Savidge tried a snare trap designed by her husband. "Tom is really interested in technology, so this became a challenge for both of us," she said. The snare trap was made out of a triangle of hardware cloth, a flexible, screenlike building material. "Inside of it a quail was tethered at the back part of the trap and there was a treadle and when you pressed down the treadle a monofilament line would close," she explained. The line would jerk the trap closed, wedging the prey in the folds of the hardware cloth.

"It worked beautifully with a human arm. You put your arm in and as it hit the treadle, it would snap down on you. But it turned out to have quite a few technical difficulties."

Indeed, the flying snare proved to be a danger to the bait and a potential catapult for the intended victim. "It was a disaster," Savidge conceded.

The fact that each of these experiments required a live, store-bought bird for bait and would, if she ever found one that worked, demand many more also troubled her. "I always hesitate with the animal rights movement. I did have to sacrifice some birds. Logistically, it was too difficult to build a trap that would preserve the birds. Scientists are very sensitive about that now. But I was faced with watching a whole avifauna going extinct, month after month. There was a sense of panic, and it was my job to find out why. I was shouldering a pretty big burden, at least I felt I was."

Savidge's dreams were now haunted by snakes. "She'd have nightmares about the snake," her husband recounted. "Once, she got up in her sleep and was walking around the bedroom brushing the walls with her hands. I asked, 'Julie what are you doing?' and she said, 'I've got to get these snakes off the walls.' "

Other times, he would find his wife sitting up in bed in the small hours of the morning, wide awake, sifting through the details of the snake theory and snake traps. "That's how it is when you are trying to figure things out and things aren't working," he said. "I remember that was Guam, for the first year. . . . It was a constant thing."

Savidge also wrote letters to herpetologists seeking information about snake traps. One such letter found its way through the Fish and Wildlife Service bureaucracy to Tom Fritts, a service herpetologist based in Albuquerque, New Mexico. "We got a letter asking what we knew about trapping snakes. There wasn't much known. A few people had done it. I think I sent her some plans for a funnel trap, as did some other herpetologists she wrote to," Fritts said.

By the end of 1984, Savidge had received the plans. They were helpful but didn't solve all of her problems. Fritts had sent her a plan for a ground-level, passive, unbaited funnel trap made of metal. She needed a trap that could be baited with a bird and hung in a tree.

In January 1984, one of her DAWR colleagues, a fisheries biologist, suggested making a funnel trap out of a tough plastic webbing, contending that snakes didn't like metal. Savidge was not convinced that metal wouldn't work, but she was ready to try something else anyway. The stiff metal hardware cloth had left her hands filled with cuts.

Much of Savidge's trap building took place after dinner on the narrow porch of her apartment house—a two-story orange stucco building on a back road in Toto. There, in the shade of mango, avocado, and palm trees, she and Tom would work with webbing, scissors, and knives. "It was really sort of pleasant," Seibert said, "and if you needed a break you could just pluck a mango."

By February, the two had fashioned a new funnel trap of plastic webbing, designed to hold a live quail as bait in the middle. Savidge took her handiwork to the DAWR lab and put it in a large cage with Big Guy. She waited. A day passed. Another day passed. Big Guy wouldn't enter the trap.

For the next few weeks, Savidge was distracted from her traps by the arrival of John Groves, the curator of reptiles at the

Philadelphia Zoo. All the commotion caused by the arrival of rails and kingfishers at the Penrose Laboratory, where Shelton occupied the neighboring office, had piqued his curiosity.

Along with the arrival of the birds had come some Polaroid snapshots of the snake, which Shelton had said was a prime suspect in the massacre of the birds. He introduced the animal as "the Philippine rat snake."

"I took one look at the photo and I knew it was the brown tree snake," Groves remembered. What the picture told him was that a nocturnal, tree-dwelling reptile from Melanesia had found its way a thousand miles northeast to Micronesia.

True, the snake did eat birds, but there were plenty of birds on Papua New Guinea, Australia, and the Indonesian islands. In fact, there were many more birds than tree snakes. "It isn't all that common," Groves said. "You tend to find it around villages."

And the villagers didn't think much of it. In a corner of the world populated with dazzling and deadly serpents, such as the tiger snake and death adder, the drab brown tree snake was seen more as a minor pest than as a peril. It was known as "the slow snake."

"I really didn't believe it could be causing that much trouble on Guam. Even on the plane out there, I remember thinking it had to be the monitor or something else," Groves said.

It had been roughly six months since Savidge had first announced that the brown tree snake was wiping out Guam's birds, and Groves was the first bona fide snake expert to turn up on the island. More than anything else, Savidge hoped he would know how to build a snake trap. He didn't.

"Although John hadn't done much trapping, he did tell me how to take care of snakes in captivity, and it was nice to talk to him about the ecology and biology of snakes," Savidge said. Of course, even Groves was to learn a thing or two. When Savidge introduced him to Big Guy and Claudia, she also showed him the hamburger balls she fed them.

Groves, citing the conventional wisdom, told her that snakes won't eat dead, let alone processed, food. Savidge was surprised. She popped a couple of hamburger balls into the cage. The snakes devoured them. It was Groves's turn to be astonished. It would not be the last time that the brown tree snake confounded the experts and defied established reptilian dogma.

Groves spent his two weeks on Guam hiking through the forests—particularly at night—and exploring the pockmarked cliff lines. To date, little of this nighttime field work had been done. But Groves, more aware of the snake's nocturnal nature, knew the small hours of the morning were the best time to study the animal. What he found amazed him. "It *was* the snake," he said. The densities of brown tree snakes were, Groves estimated, so high that it wasn't surprising that the birds and probably everything else in the forest the animal could eat was being wiped out.

"They asked me how many snakes were out there," Groves recalled, "and I said well here's my guesstimate—a million and a half—prove me wrong." That would have meant there were more than ten times as many snakes on Guam as people. Then Groves got on an airplane and flew back to Philadelphia much impressed by what he had seen. When he arrived home, he sat down and wrote the first herpetological report on Guam's "snake problem."

He was reluctant to blame the entire disappearance of the island's birds on the snake, saying other past factors such as bombing during World War II, increased air traffic during the war in Vietnam, pesticides, and habitat loss might have contributed to depressing native populations. Still, he concluded, "the introduction of the brown tree snake had dramatic effects." If it had not been the first cause, it was certainly the last.

"A snake," he said, "has never been involved in anything like this before. We've never seen a reptile cause so much ecological havoc."

The visit left Savidge with some comfort. Perhaps Douglas Pratt hadn't seen snakes and didn't believe her, but the first genuine herpetologist to visit the island had seen legions of brown tree snakes, and he believed her.

Savidge simply had to get a trap to work. She fiddled with her latest prototype. On April 5, Muna placed three of the funnel traps into the large cage in front of the DAWR lab that Big Guy called home. One cage had a sparrow in it. One had a quail.

One had a chick. Big Guy ate the sparrow and then the quail. "Big Guy," Savidge said, "got pretty big."

Now there was hope, Savidge had a trap that a snake would enter. On April 17, Savidge and Muna placed seven funnel traps out in the field. Some were hung from trees on the metal wires, some placed on stumps, and some left in the brush. After two nights one of the birds was gone. The next night another bird was gone and a snake was in the trap. By the fourth night, all seven birds had been eaten. "This is really hard to believe," Savidge wrote in her field notebook; "all the birds are gone."

"I was so excited because it worked," Savidge said. "These birds got nailed within four days. . . . I finally had something that worked. . . . There was a tremendous sense of relief."

The time to test hypotheses had come. But in addition to her three original suppositions, John Groves had given her something else to ponder—the densities of snakes. Through trapping, Savidge believed she would be able to calculate a better estimate of the number of snakes hiding out there in the night. Could Groves be right? Could there be millions?

Armed with her new weapon, Savidge began stalking the tree snake across Guam. She set out traps on Cocos Island, where birds were still abundant. She placed traps at the Naval Magazine, where the snake had made its beachhead more than thirty years before. She put traps in the open savannas, where black francolins, yellow bitterns, and blue-breasted quail could still be found. Traps were set in the scrubby second-growth woods along the perimeter of Northwest Field and the Andersen Air Force Base flight line, and finally in the virgin forest at Ritidian Point. In all, over the next year she would create seven different trap lines around the island.

These traps were not just casually plunked down. With Muna's help, Savidge constructed transects, similar to those used by John Engbring to count birds, with fifteen to thirty-two traps. Each device was spaced precisely 160 meters (525 feet) from the next one. A blue-breasted quail was placed in each trap and provided with seed and water every day. Every day, that is, that the bird was still there.

Going from cage to cage at Northwest Field one morning Savidge made these notes: "Snake caught. Bird gone, snake caught. Bird gone, snake. Bird killed by snake. Bird gone, snake. Bird gone, snake. Snake caught. Bird gone, snake." Looking over the results, Savidge said, "In seven nights there was 100 percent predation."

Now the hypotheses could be tested against the trapping data. Savidge had reasoned that areas that still had birds would have lower rates of predation and areas where the birds had vanished would have higher rates.

"Sure enough, on Cocos Island, where the birds were doing well, there was virtually no predation," Savidge said. Down at the Naval Communications Station, 75 percent of the traps were hit within four nights by snakes. "We put them down at Ritidian. Monitors were a problem. So we had to take the traps down during the day. But even doing that 65 percent of the birds were hit by snakes. I spent several nights [in the Ritidian forests and] we saw snakes on trees, snakes moving through the forest" she said.

A year before, Savidge had sat paralyzed by Guam time, unable to collect the data she needed. Now, she found herself on a treadmill. She was out at night, like Groves, collecting and observing snakes. She was still trying to mist net birds in the early morning, and she was checking and servicing her traps during the day. "I kept really crazy hours," she said, "and that was the nice thing about DAWR, they let you work when you had to . . . there was no punching a clock."

To address her second hypothesis—that other types of prey should also be declining—Savidge had to establish what the populations of mice, rats, and shrews had been before the snake and what they were now. Luckily, R. H. Baker had, in 1946, published a monograph entitled "A Study of Rodent Populations on Guam, Marianas Islands," and K. R. Barbehenn had written two papers on rats and shrews in Micronesia in 1974.

With the mice and mammal counts of Baker and Barbehenn in hand, Savidge set out to trap these furry little animals using fresh coconut and peanut butter. She replicated the two researchers' experiments and discovered that while Baker had trapped 30.4 to 68.5 rodents per hectare at Santa Rosa in 1946, she could only get 2 per hectare.

Barbehenn had recorded densities of shrews of 15.3 per

hectare in 1964. Savidge now found a 94 percent decline. In fact, on some parts of the island, it was difficult to find any small mammals at all.

In 1981, Chris Grue had trapped shrews as part of his pesticide study and had found them in most of the second growth and mixed forest of northern Guam—in the same places that Engbring found birds. Now both birds and rodents were gone.

Savidge was well on her way to proving that the snake was a voracious predator. But that didn't mean there wasn't also a disease, and the Fish and Wildlife Service Honolulu office still seemed reluctant to give up on the disease hypothesis.

"The Fish and Wildlife Service in Honolulu was convinced that this was an avian disease problem," Bob Beck said. "In fact, we were essentially forced to do the disease work by them because they were convinced it was disease."

Julie called Lou Sileo and urged him to come back to Guam. Sileo agreed, and this time he planned to return with a multitude of birds—from Iowa, Wisconsin, and Maryland—for an unparalleled experiment, an experiment aimed expressly at finding that "Guam disease"—if it existed.

July 1984

Tom Fritts awoke in a room at the Downtowner Motel, overlooking Marine Drive, Guam's main drag. He was just a bit jet-lagged from the inescapable red-eye flight to the island.

Fritts had come in search of snakes. He had first read about Guam's "alleged" snake problem a couple of years earlier, in an article buried inside some newspaper. "At that time, I worked for the National Fish and Wildlife Laboratory in Albuquerque, which was a research institute within the U.S. Fish and Wildlife Service. I wrote a memo to my boss and said, 'This sounds like a classic case for a Fish and Wildlife Service herpetologist with experience working on tropical snakes. . . . If this is real.' "

But nothing came of the memo. Then Julie Savidge wrote to the director of the Fish and Wildlife Service seeking information on snake traps. Her inquiry had found its way to Fritts. Savidge's work was enough for him to wangle a plane ticket to Micronesia.

He was the second herpetologist lured to the island by the prospect of this unique reptilian phenomenon. Fritts had studied alligators in the Louisiana bayous, giant tortoises on the Galapagos Islands, and lizards in Mexico's Baja desert. But he really didn't expect to be adding the brown tree snake to his curriculum vitae. "I went to Guam the first time as a doubting Thomas," Fritts said. "I thought, 'If it's really true there have to be four or five criteria that are really obvious that nobody has mentioned yet.' "

And so, on the morning of July 5, Fritts left the Downtowner to get his first glimpse of Guam. Marine Drive runs along the island's western shore from the crescent of fancy hotels at Tumon Bay in the north to the U.S. Navy facilities in the south.

The Downtowner sits on the drive, just where strip development gives way to the strip joints and beer halls that are general issue around American military installations. Across the four lanes of traffic are the stadium where Guam's semiprofessional baseball teams play, a park, and a marina.

There is also a big, muscular statue of Chief Quipuha, who aided the Jesuits in their seventeenth-century campaign to spread Christianity to the Marianas Islands. The figure is larger than life, with taut, angular muscles and a flat stomach. Clearly, Quipuha lived in the time before beer and Spam became island staples.

These days a lot of Chamorros think it would be more fitting to have erected a statue of Chief Matapan, the slayer of Padre Diego Luis de Sanvitores—the Church's primary apostle and zealot on Guam.

Christianity had been a hard sell to the aboriginal Chamorros. Although they believed there was an immortal spirit, good and evil had no part in determining the soul's fate. As far as the Chamorros were concerned goodness was its own reward in this life.

They had no single great god, but worshipped the spirits of deceased kin—known as the *aniti*. They also believed that all humankind had been created from a precious primal substance found in a single cliff on Guam.

With no sense of guilt and their own version of creation, many Chamorros saw no need to sign on to the religion of the bearded men in the big ships. But Sanvitores was not to be deterred and when necessary he used Spanish soldiers to carry out forced baptisms.

Unfortunately, the padre didn't have his armed altar boys around when, against Matapan's wishes, he baptized the chief's son. Sanvitores's murder sparked years of bloody repression of the Chamorros by the Spanish military.

Sanvitores, while the most prominent missionary, was neither the first nor last priest killed—a dozen more followed. It finally got so bad that the dismayed Spanish crown revoked the Jesuits' commission for the Marianas and gave the Dominicans a crack at converting the natives. Something must have worked, for today Guam is a deeply Catholic community, although the passion for fiesta and bright flowers may speak to those original tropical roots. Still, more and more young Chamorros look fondly upon old Matapan.

Fritts stepped out into the eye-scrunching sunlight with Marine Drive, Chief Quipuha, the baseball stadium, and the glittering Pacific spread before him. But he saw none of it. His eyes were riveted on the asphalt at his feet.

"There was a dead snake right in the road," Fritts said. "I got up and walked out of my hotel and found a dead snake within 100 yards . . . in one of the most urbanized areas of Guam. I walked on down and next to a used car lot I found another dead snake." Two reptile road kills virtually on Fritts's doorstep.

He borrowed a bicycle and pedaled about twenty miles through northern Guam. "You can see a lot more things on a bicycle," he said, "and sure enough, that one day I saw eight or ten snakes dead on the road."

He went out at night to check fence lines and found himself in the custody of the military police, who wanted to know what he was doing prowling around the perimeter of a U.S. Air Force installation. But again he found snakes.

"Basically, nearly every place I looked, I could find evidence of snakes. . . . Even if I couldn't find them dripping off the trees like icicles, they were there," he said.

And, of course, Fritts paid a visit to the DAWR, where his arrival was viewed warily by both the agency and Savidge. Unlike Groves, who had come as an adviser to the DAWR, Fritts had his own brief. "I had been sent out to evaluate the situation and see if it was real, see what could be offered up as information to bolster the case or tear it down, and determine what the [Fish and Wildlife] Service should be doing or encouraging the DAWR to do," Fritts said.

The DAWR already felt it was being told how to run its endangered species program and now division officials were chary that the federal government was going to tell them what to do about the snake as well.

Savidge worried about the fate of her data and her dissertation. "Julie was very proprietary about her data," said Gary Wiles, the DAWR biologist responsible for tracking the bat populations, "and she had very specific ideas about how things ought to be done."

Savidge said her thesis committee had warned that her material simply could not be published before her doctoral work was completed. Fritts, however, had his mission, and with a measure of diplomacy and thick-skinned indifference he was determined to pursue his assignment.

Personalities aren't supposed to figure into science but they do, and so, for a time, the two most knowledgeable biologists on the subject of Guam's brown tree snake exchanged information in a professional but strained manner, and circled each other warily on the small island.

Neither has much to say about their working relationship. "Tom and I did not work all that closely together," Savidge said. "He came out with an agenda, so we didn't spend all that much time together. But he seemed receptive to the idea of the snake as a problem." Fritts would only say, "She was busy working on her dissertation, whatever."

But Fritts, like Groves and Savidge, was becoming a believer. "One of the things I went looking for was the idea that if the snakes were really doing this, there had got to be a gob of them, and indeed I saw evidence that there were a gob of snakes. No way of making a quantitative estimate beyond a gob," he said.

It also struck Fritts that if there were that many snakes on the island they must be showing up in places other than birds' nests. He mentioned this thought at the DAWR and Bob Anderson dug into a file and produced some data on snake-induced electrical power outages that he and Savidge had compiled. "I looked at it and said, 'Holy cow, you've got to use this,' " Fritts said.

The data showed a pronounced increase in electrical outages around the island caused by snakes hitting the lines of the Guam Power Authority. In 1978, there had been ten outages. By 1982, the number had climbed to sixty-five. "That told us Guam had many more snakes than anyone ever thought, enough to cause uncommon electrical problems," Fritts said. "That was a tough one to imagine, but it was true. Basically, it came to finding snakes every place. That is a very extreme situation in itself."

Fritts began to review what was and was not known about "this obscure snake from an obscure part of the world." The brown tree snake is a member of the class Reptilia, order Squamata, suborder Serpentes, family Colubridae, subfamily Colubrinae, genus *Boiga*, species *irregularis*.

This was a beginning, but not a great one. There are a total of 2,700 known species of snakes, and 2,200 of them are lumped into the sprawling, unwieldy, little-studied Colubridae family.

In general, snakes have been less studied than many other orders of animals. But even among those that have been scrutinized, scientists have tended to be drawn to the evolutionarily "more advanced" families—Elapidae (cobras), and Viperidae (vipers), or to the primitive but giant Boidae (boas).

It is the more complex behavior, physiology, size, and economic value of these snakes that make them more interesting to study. For example, the highly developed venom delivery systems of vipers have attracted the curiosity of a number of scientists. Adding to the interest is the fact that these snakes are among the most deadly and, hence, the greatest threats to humans. So in an area like Melanesia, where colubrids, elapids, boids, and vipers all live together, the evolutionarily primitive colubrids lose out, as they do most places, in the scientific popularity contest.

The brown tree snake was part of this large, rambling family, whose members are described as "typical snakes." To be a member of the colubrid club a snake must have a loose facial structure, have few head scales, and lack both a left lung and front fangs that inject venom. These requirements are so general that the club includes everything from the garter snake to the venomous African boomslang. Not only do the family members differ substantially, they live literally all over the map—from Australia to Canada, from Argentina to China. In fact, there are colubrids in jungles, forests, deserts, and temperate climes. They exist everywhere there are snakes.

There are some snakes, like rattlesnakes, that have been well studied by virtue of both being dangerous and living in relatively close contact with man.

But within the vast snake family, the brown tree snake is a member of the biggest subfamily and one of the largest genera—*Boiga.* Just another snake in the crowd.

Boiga are found only in Africa, Asia, and Australia. This particular *Boiga* is known as *irregularis* because of the irregular size and number of scales on its back. Scale patterns are as distinctive as fingerprints, and so counting scales is one of the primary ways that herpetologists identify snakes.

So what is known about *Boiga irregularis?* It is a native of a large number of islands in Indo-Malaysia, which extends from Sulawesi in eastern Indonesia through New Guinea to the Solomon Islands and into the wettest areas of northern Australia. Generally, the snake grows to about six and a half feet but remains thin as a rod, weighing at most about two pounds. Its color runs from drab brown to drab olive, apparently depending on its location.

It is known for eating a wide variety of prey, from lizards to birds' eggs to chickens to cats. It is not uncommon to find a brown snake stuck in a chicken coop in the morning because it is so bloated on chicken that it can't pass back through the narrow mesh.

Just looking at it—slender, with a barely detectable ridge along the base of its ribs, and a prehensile tail for grasping—it is clear that this is an arboreal snake. Its large eyes with elliptical pupils tell of its adaptation for hunting at night.

Not much is known about the animal's sex life. It has been reported that the female produces four to twelve oblong, leathery-shelled eggs, each about forty-five millimeters long. It is also thought that the female, like a number of other snake species, can store sperm and produce fertile eggs for years after mating.

In 1973, John Groves wrote a brief paper about a similar species, an Asian mangrove snake, *Boiga dendrophilia*, which came to the Philadelphia Zoo in October 1967 and was still dropping fertile clutches in January 1969, even though the zoo didn't have a male snake.

This ability to hold sperm for years and the fact that snakes can live for months with no food and only a little water would make them, Fritts realized, highly effective at colonizing new territories.

Although the snake is basically shy and secretive, it will fight fearlessly when startled or trapped. It will lunge and bite repeatedly and lash with its tail. Its fangs, however, are at the back of its mouth, and unlike a rattlesnake, it delivers its venom with less than hypodermic efficiency. Instead, the venom is drawn down a groove on the side of the tooth and into the wound. There are reported cases in which venom caused localized swelling and itching, but it is not deemed dangerous to humans.

And that's it. Pretty much the entire sum of knowledge in the world on the brown tree snake. Fritts wasn't surprised. Even when it comes to the better-known snakes, like cobras and boa constrictors, there is precious little information. "There may be mentions of a species in forty books," he said, "but in thirty-eight it's size and eating habits and the other two will have something else—like mating habits—and that's all we know.

"There's been a lot of interest in birds and mammals, ever since science began . . . ," Fritts said. "Man has put a lot of emphasis on them. But reptiles are just now coming into their own. We are realizing that reptiles are important from an ecological point of view and from the point of view of being endangered species. So we are playing catch-up ball."

Of course, dealing with an animal like the snake, which is wary, secretive, and usually a loner, does present the biologist with its own set of problems. Naturalists delight in working with birds whose courting, mating, and rearing of young are all highly visible. Biologists can easily trap and track mammals, enabling the scientists to follow entire populations of animals. But snakes are few and far between and difficult to catch or trap—except on Guam.

Since the snake was emerging as the prime culprit in Guam's bird massacre, the DAWR was anxious to find out more about the animal. Where did it live? How did it mate? How how often did it have clutches of eggs? Bob Anderson wanted answers to those questions. One solution was to follow snakes. Obviously, they couldn't be tracked like lions or deer. The only way to do it was to "bug" a snake with a transmitter and radio track it.

That could be done. The question was, who would do it?

Savidge already had her hands full trying to net birds for her disease study and trap snakes for her predation analysis. Beck was working full-time bagging kingfishers and rails and caring for those in captivity.

Somebody else would have to do the work. Gary Wiles, the division's bat biologist, was that somebody. "Harry Kami, our boss at the time, and Bob Anderson were pressuring to get going on snakes more and more," Wiles said. "When I was at Purdue University I had done radio tracking. . . . I had some radio-tracking experience, some telemetry experience. So I said, 'Okay, I should do something to help figure out this snake.' "

But his heart really wasn't in it. "To be quite frank, I didn't have the desire to do it. I wanted to be doing my bat work. By that time I was going full bore on the bats, and suddenly this was thrown in my lap. I was aware of what a good telemetry study requires. It requires full time in the field. It requires day and night work. It is very time consuming. The office has always been one of these deals where you get all this work dumped on you and you do what you can. But you never expect to spend forty hours a week in the field, let alone sixty or eighty."

Wiles had come to Guam in 1981 after completing a master's degree in biology at Purdue, where he studied Indiana white-tail deer. Deer are mammals. Bats are mammals. Wiles became the DAWR's bat biologist. On a small island you've got to make do. When Bob Anderson had arrived on Guam back in 1978, the island's bat population had dwindled to less than 100 Marianas fruit bats. Five years later, the population had reached about 800 and Wiles was anxious to document the bats' recovery. But along came the brown tree snake.

Wiles wrote to several companies specializing in telemetry equipment to see what kinds of transmitters were available. The devices ranged from the size of a pinkie nail to one as large as a man's thumb. This presented Wiles with his first dilemma. A small radio was less intrusive but it also didn't transmit for very long. The general rule is that it takes an animal a couple of weeks to get back to its normal routine, and then add to that the fact that a cold-blooded reptile might spend a week sleeping under a rock and it was distinctly possible that absolutely nothing could be learned during the month a tiny radio was transmitting.

So, he opted for the bigger, thumb-sized radio and its five-inch wire antenna. "They are designed for big animals," Wiles conceded. "They work great on deer."

Since the radio had to be embedded in the animal's abdomen, the larger transmitter limited the study to only the largest of the brown tree snakes on Guam. But they were not found in the forests, where biologists were most interested in the animal's behavior. Instead, they inhabited urban areas, where snakes apparently could find more food and grow larger.

From the outset, Wiles realized that using large snakes seriously limited the study. "At the time they made up a small proportion of the population, maybe less than 5 percent," he said. "So by studying the very big guys, you are getting a very biased view of how the population is behaving." Still, it was a start.

To get those snakes, Wiles and DAWR technicians first tried going to the forests, without success. Then they started hunting in developed areas. "What we were doing was catching snakes around town and then putting transmitters in them and releasing them up at Tarague [near Ritidian Point], thinking that would be somewhat natural."

Wiles would anesthetize each snake, make a slit in the abdomen, insert the transmitter in the gut area under the rib cage, and sew it back up. That was no problem. The problem was the five-inch wire antenna. "I could have sewed up the transmitter and left the antenna out, which seemed like it would very likely get caught on something and the whole thing would get ripped out," Wiles explained.

"So what I ended up doing was inserting a tube down underneath the skin of the snake and making a little incision—this is based on the literature of other researchers—where the end of the tube was, then running the antenna down the tube and pulling the tube out . . . and leaving the antenna in place.

"That was traumatic to the snakes. You could see it. Even though they were under anesthetic, they'd squirm when you rammed that tube down under their skin like that. It was a rude thing to do to an animal."

The alternatives, however, were to use a smaller radio and risk getting no data, to leave the antenna out, which would surely injure the animal, or to abandon the radio-tracking study.

The snake was kept for two days after the surgery to allow it to recover and then it was taken to Tarague Basin. Wiles hoped for "natural" behavior on the part of the wired snakes. It didn't work out that way.

The first snake was set free while Fritts was on Guam. Indeed, Fritts joined Wiles in the field the day after the animal had been released.

It had made its way to a clump of pandanus trees—whose maze of roots offered a perfect hiding place for the snake. Fritts approached the pandanus from one side, Wiles from the other. Fritts spotted the monitor first. The large lizard was dragging the snake's carcass out of the clump and into the woods. So much for radio tracking that snake.

Wiles did not have much better luck with the next two subjects. "They were moving around constantly from one day to the next, but the hiding spots where they were sleeping were just too obvious," Wiles said. "You could see a snake sleeping on the ground under a rock. He was sprawled out and part of his body was sticking out on the ground. It was like he didn't know where he was. He seemed totally lost," Wiles said.

"Another time I saw one sleeping in a low tree, but out in the open. I was seeing the radio-tracked snake much more frequently than I would normally see a snake out in the wild. So I guess the feeling was that things weren't natural," he said.

The fourth snake also met with an untimely end. Wiles had been tracking the animal for nearly two weeks. Each day he would walk the area where the snake had been released and determine its whereabouts. Then one day he went into the area and "smelled something dead." He looked around and found a half-eaten snake. He looked a little more and found the radio transmitter. On the radio's wax coating were the teeth marks of a monitor lizard."

Odd behavior and a 50-percent mortality within two weeks. "It told me this was no good for the animals," Wiles said. "We decided to start again from scratch."

Besides, there were Wiles's bats. In fact, he had recently noticed something curious about the recovering bat population. While there were adults and babies to be seen, he hardly ever found juvenile bats anymore. It seemed that babies were born, were nurtured, and then disappeared. "Where are they going?" he wondered.

The brown tree snake was eating birds and causing power outages, but Fritts was slowly coming to the realization that those weren't the only problems it was causing. Perhaps they weren't even the most serious. As he traveled around the island, visiting the various military installations, he realized that the snake posed another threat.

He saw ships coming and going, airplanes landing and taking off, military transports of troops and equipment using Guam as a way station, and he thought, "If a brown tree snake accidentally came to Guam in military shipments, why couldn't it accidentally get shipped someplace else?"

Indeed, the densities of the snake were so high in some areas that such a prospect seemed likely. "I talked to Seabees, who move large amounts of equipment, and asked if a snake could get in, and they said sure," Fritts said. Interviews with navy and air force personnel revealed that snakes had already been found in bombers, helicopters, aviation fuel trucks, and drainage pipes leading from aircraft parking pads, and in the vicinity of cargo dispatch areas.

In a report Fritts wrote upon his return to Albuquerque, he noted "several aspects of the biology of this snake" that make it an ideal stowaway. It is a nocturnal animal that moves through the dark and seeks a dark, secluded hiding place during the day.

Its natural haunts include hollow trees, crowns of palms, rocky cliffs, caves, and other natural features that provide dense shade and moderate temperatures. But when it occurs in high densities in areas close to human settlements, drainage pipes, rafters of warehouses, vehicles, cargo crates, and airplane wheel wells can all make excellent hiding places too.

In addition to being a military way station, Guam was the central commercial transit point for all of Micronesia and points in the South Pacific.

There were hundreds of islands spread across the Pacific—all the way to Hawaii—that harbored unique bird populations and no snakes. These islands would fare no better than Guam if the tree snake reached them.

No, the crucial problem, as Tom Fritts saw it, wasn't that Guam was being overrun by an exotic snake. The risk was that Guam's problem could metastasize like cancer and spread across the Pacific. On July 27, Fritts left Guam. But he would be back.

Chapter

10

July 1984

The day after Tom Fritts left Guam, Lou Sileo arrived at the Agana airport to begin a major disease study. This time he came with eight boxes of equipment, an ice cooler, a tank of liquid nitrogen, and sixty-four canaries in six cages. There was no one to meet him. Nobody expected him. It was a Sunday afternoon.

You don't want to be taking a taxi with sixty-four canaries, Sileo thought. He started making phone calls and found Bob Anderson at home. An hour later, Anderson and Savidge arrived with a couple of cars, into which they loaded the canaries and equipment, and headed to the DAWR compound. Apparently, the international date line had proved too daunting for the secretaries at the National Wildlife Health Research Center: they had advised the DAWR that Sileo would be arriving on the 29th, not the 28th.

Thus began one of the most ambitious "sentinel studies" ever done. The goal of the exercise was to take a bunch of clean, healthy birds, birds carefully bred (Sileo called them "disease naive"), expose them to whatever was on the island, and see if they'd get sick.

There had been sentinel studies on Hawaii with Laysan finches to see if they'd catch avian malaria. There had been studies in Maryland with bobwhite quail to test for equine encephalitis and in Michigan with Canadian goslings and mallards that caught leucocytozoonosis. There had been studies with chickens and with turkey poults. In just about all the experiments, the birds had contracted the suspected disease within a week.

This study would involve nearly 300 birds from four different species and run for more than a month. If that many birds—most of them never exposed to any germs in their lives—totally vulnerable to whatever was out there didn't catch whatever was out there, it might safely be said there was nothing out there. It was to be the final, grand assault on the disease hypothesis.

One of the principal arguments against the snake theory was that a snake simply had never extirpated an avifauna. But Sileo said: "What I knew and what none of them knew—I was the first pathologist there, and they were all biologists—was that there was no example, except perhaps avian malaria in Hawaii, of a disease extirpating species. There have been some implications of trichomoniasis with the passenger pigeon and coccidiosis with the heath hen, but that stuff is speculative. So really there is no more support for the disease hypothesis than the snake hypothesis.

"There is no disease process that is going to wipe out an entire avian community. Diseases just don't work that way," he explained. There will always be some individuals resistant to disease and some species that are stronger than others. "Maybe disease will knock down a species, but extirpate a community?" Sileo said skeptically. "Anyway, theoretically, if a disease is wiping out an entire community we ought to find something."

Since Savidge had arrived on Guam, she had amassed 112 dead bird carcasses to be necropsied, 260 blood smears, 212 swabs. "We never found anything helpful. We isolated the occasional organism and made the occasional diagnosis—emaciation, trauma, dermatitis, pulmonary hemorrhage. We did find things, but we didn't find anything that would extirpate an avifauna," Sileo said.

Sileo himself had screened all the bloodstains, all the bacteriology, virology, and necropsy reports, and found nothing that offered a clue. There were seventeen years of experience and a host of tests telling him that there was no disease on Guam.

When he had returned to Madison from his first trip to the island, Sileo had strongly argued, spurred by the analysis of Savidge and Beck, that the birds were disappearing so quickly that if they didn't do a sentinel study immediately, they would lose any opportunity to do one at all. But it was late in the federal fiscal year and the extra dollars simply could not be found.

"So I wrote it off," he said, "I said too bad the birds on Guam are gone. We'll never know why. And I started to get on to other things." When Julie Savidge called him the following spring to urge that they press ahead with the sentinel study, Sileo was surprised. He sheepishly went back to his superiors and told them that the study he had insisted absolutely, positively had to be done last year—the one-time opportunity—could indeed still be done, if they hurried.

He doubted that the service would spring for it. But Guam's problems had by this time gained "political momentum," Sileo was surprised to discover. There had been the front-page flap with the U.S. Air Force over the rail habitat, and zoos were now interested in participating in captive breeding programs, although they were worried about taking birds from Guam.

"These zoo people would call me and ask if it was safe to take these critters, and I'd say we didn't know yet. The zoos were a little bit edgy about accepting critters from where they have a mysterious Guam disease that's wiping out a whole avifauna," Sileo said.

The bureaucracy wanted answers now, and they were willing to pay for them. "It seems that in so much of the work I've done for the past ten years there is a basic biological problem, but the main reason the work is done is for the political implications," Sileo observed.

So he was back on Guam with canaries and gear to embark on this great experiment. After placing the birds in a special mosquito-free room at the DAWR, a weary Sileo headed back to Savidge's apartment, where he was going to spend the night.

It had been a long day and he was looking forward to bed, but Savidge and her husband, Tom Seibert, wanted to talk. "I remember getting a real grilling, mostly from Tom, which irritated me a little bit; what the hell did he have to do with any of this? I've learned since, in fact, that he'd had quite a lot to do with it. I was jet-lagged and tired. It was a tough trip. It was a

Sunday. I was in their home. That first trip I had gotten along so well with Julie. This time it was different and they really gave me a grilling on some of the decisions I had made. That's when the friction started," Sileo said.

This was not just Sileo's experiment. It was Savidge's also and it was crucial to her work, her dissertation, and her ability to prove that the snake was the primary source of Guam's avian crisis. She wanted to be sure everything was just right and that the data would be hers.

Savidge and Seibert questioned Sileo on the "reference dates"—that is, the dates that they would use to start counting exposure for the animals. They challenged him on his decision to use a drug to suppress the immune systems of some of the test animals.

"There was tremendous resistance to that because immuno-suppressing them was hazardous. We didn't know much about dexamethasone—that was the drug. At that time there wasn't very much known about dexamethasone in birds," Sileo said.

Savidge's main concern was that if the dexamethasone killed the birds or caused them to get some common ailment, it would defeat the goal of the sentinel study. Similarly something as trivial as the reference dates could either heighten or reduce the calculated exposure to the mysterious Guam disease. Sileo believed that immunosuppressing would give the birds the greatest possible opportunity to catch the Guam disease.

He tried to be diplomatic. "I had to adamantly, strongly convince her that she would be lead author on this stuff," Sileo said. "We just wanted to get our study done and find out what was going on and give the answers to our agencies out there so they could get on with their work. The data were hers. She would be lead author right up through publication," he said.

But when it came to the research design, picking reference dates and deciding how to handle the "test units," as he called the birds, this was Sileo's bailiwick. He was the senior scientist and he wanted it his way. "I had a lot of trouble convincing other people we ought to do this," he said.

He did, however, convince DAWR chief Harry Kami and that settled the dispute. The birds would be injected with dexamethasone. But that was just the first argument. The sentinel study was to run about a month. Clearly, the next thirty-one days were not going to be easy.

Things got a whole lot harder two days later. The research plan was to take a total of 278 birds—chickens, quail, canaries, and white-eyes—and place them in the northern limestone forests near the remaining populations of birds and at the DAWR lab for a control site. The study was being done in the middle of the rainy season, when mosquito populations were highest on the island. One test site was at Potts Junction, at the top of the limestone cliffs, on air force land. The other was down in the Ritidian forest.

Beck had surveyed the areas and established that there were indeed still low densities of birds in the woods. They were there, but they had dropped to such reduced levels that just before the sentinel study began, Savidge had abandoned any further efforts to net forest birds. The odds of catching one had become just too low.

Muna had already macheted pandanus, hibiscus, and ink-berry bushes from the two test sites to create clearings and constructed a complex of four large cages, which really looked like a small shantytown.

That may not be a fair description, because some of those cages were built better than many a house. All the cages were designed to thwart feral pigs, monitors, rats, and snakes from getting at the test birds. The cage that held the exposed canaries and white-eyes was eight feet long by eight feet wide by six feet tall. It was covered with hardware cloth. Inside, suspended by ropes from the ceiling and about three and a half feet from the ground, was a platform with another hardware cloth cage, twenty-two inches by twenty-six inches by eighteen inches.

The chickens and quail would be kept in a couple of simple cages, two and a half feet by two and a half feet and four feet tall. They stood on legs that had been smeared with Tangle-foot—a sticky resin product designed to deter insects and snakes from crawling into the cages.

And then there was the "control" cage, the Fort Knox of the experiment. This structure was built to exclude flying insects as well as the other predators. It consisted of an outer structure, roughly eight feet by twelve feet and six feet tall. It was covered with hardware cloth on the outside and a layer of window screen on the inside. Within this was a second smaller structure, about five feet by eight feet by five feet, that was covered with mosquito netting.

Both cages had hinged doors and gaskets around the edges to insure a tight fit. There were also little anterooms of netting to pass through before entering both the outer and inner chamber.

The plan was to place some of the test animals in the relatively open cages and others in the control cages and see what would happen. Would the chickens, quail, canaries, and white-eyes left out in the open come down with some new and devastating disease?

A month earlier, Savidge had netted sixty-four white-eyes on Saipan for the experiment and had kept them in the mosquito-free room at the DAWR lab. Sileo had purchased his sixty-four canaries at the Discount Pet Center in Madison and had flown with them in mosquito-proof cages to Guam.

The remaining birds—chickens from a lab in Iowa and quail from a farm in Maryland—were scheduled to arrive with Diana Berger, the young veterinarian whom Sileo had hired to help care for the animals during the study.

On August 2, Sileo and Savidge were up at 6 A.M. They went to the lab and gave the first dexamethasone injections to some of the birds. They examined all the canaries and recorded their weight and sex. They packed supplies and then they headed to the airport to meet Berger and 150 quails and chickens.

The Continental Airlines flight touched down a little after 7 A.M., but it wasn't until nearly nine o'clock that the cages were rolled out into the cargo depot down the road from the passenger terminal. When Sileo and Berger checked the cages they were grasped by cold despair. They were looking at a whole mess of frozen chicken and quail. There were 121 stone cold birds. Thirteen chickens and sixteen quail had survived, but was that enough for the study? Sileo didn't think so.

"It was very disheartening," Savidge said, "because these birds were worth a lot of money. They were completely clean—free of pathogens. They had gone through sampling and screening back in the States, and now they were dead."

While Sileo pondered his research design and Savidge her

dissertation, Berger mourned for the dead. "They were," she said, "living beings." By her own description, Berger was "an incredible humaniac" who tended to "see animals as individuals." Sileo, of course, called them test units. This would add yet another line of tension to the study.

"I knew something was wrong in the plane when we had trouble with the air conditioning for the cabin," Berger said. Her apprehension turned into "profound grief" when she looked into the cages. "I felt terrible. I thought of how those poor birds suffered," she said.

"She was the attending veterinarian," Sileo said, "and most of her charges were dead. I tried not to make her feel responsible . . . apparently it was the air conditioning on the plane. I didn't want to harass her, there was nothing she could do about it. At the same time, I was pretty disturbed and panicked. There was potentially a lot of money going down the drain here." The Fish and Wildlife Service, according to Sileo, had allocated approximately $67,500 for the study, not counting salaries or overhead, and now that was in jeopardy.

It was already August; the cages had been built, the equipment shipped. But buying "clean" chickens and quail wasn't like ordering a new set of test tubes. You couldn't purchase them off the shelf or have them made up in a day. Would it be possible to get new test animals quickly enough? Savidge was scheduled to go off-island in October and Sileo couldn't stay on Guam forever. Was there still time to do the study? Sileo was deeply pessimistic. "I thought, well, we're going to have to pull the plug," he said.

The three took the surviving birds to the DAWR lab and then went out to the field. It wasn't until nearly 11 P.M. that Berger got settled at Andersen AFB, where officers' housing had been made available for both her and Sileo. It had been a long day for everyone. A long, bad day.

Sileo was up at six the next morning to telephone the Wildlife Health Research Center in Madison. He spoke to the director, Milton Friend, who advised him not to do anything in haste. Friend urged Sileo to continue setting up the study and to evaluate the situation and meanwhile the center would try to find some new, clean birds—fast.

Sileo, Savidge, and Berger passed the day down at Ritidian getting the site ready and up at the lab caring for the 157 birds they did have. Still, they really weren't sure what they were doing. That evening Sileo and Berger went to a Korean restaurant. It was a bit like a first date—two people trying to get to know each other, a lot of chitchat, as Sileo remembers it. But the dead birds and the uncertainty of the whole venture lurked behind all the casual banter. Suddenly, Berger began to cry. "For a minute I thought she was kidding," he said, "but she was really crying." The end to another bad day on Guam.

Within another two days, however, the center had located thirty clean chickens and was trying to get them to Guam as quickly as possible. A note from a colleague was waiting for Sileo at DAWR headquarters one morning. It read:

"Hi, Lou—Hope all is going well. Things are quite chaotic here. We're having problems getting the chickens on a flight. But not to worry. I think a day without a crisis just isn't a day worth getting up for."

And so the great sentinel study lumbered forward. The plan was to start with the birds they had and rush the extra chickens into the field as soon as they arrived on Guam. The trio began moving the birds to Potts Junction and Ritidian and almost immediately the arguments began. The key data would come from blood samples periodically taken from the birds. Sileo wanted to take the blood from the jugular, where he was sure of getting a good sample of the serum, which holds the antibodies and best clues to disease. Savidge thought that would be too hard on the white-eyes.

As a compromise they tried the technique on just one white-eye. It almost died. Sileo backed off. There would be no serum from the white-eyes until the end of the project.

The next day, Savidge indicated that after the study she planned to release the white-eyes on Cocos Island—as a translocation and propagation project. "They were an endangered species," she argued.

That provoked another argument with Sileo. "I never did like that idea. One reason I didn't like it is because I'm a histopathologist. I extract my maximal, most useful data from looking at these things under a microscope, and they've got to be dead to do that," Sileo explained almost apologetically. "I wanted to put their little livers and spleens in formaldehyde."

The following morning—just three days into the study—two of the canaries injected with dexamethasone died. Necropsies were quickly done. One bird had died from trauma, the other from a liver tumor hemorrhaging. Neither appeared to have died from the "Guam disease." This sparked yet another fight over the use of dexamethasone. But Sileo prevailed. The injections would continue and about a third of all the birds would receive them.

It was remarkable that they had the energy to fight at all because just getting through the day was a grind. At its peak the study had nearly 190 birds to care for and feed. The days began with breakfast at dawn. The quail, chickens, and canaries got birdseed. The white-eyes received a mixture of nectar-protein solution, fruit, and mealworms. The white-eye trays had to be changed twice a day because it was so hot in the cages that the food spoiled.

"The nectar was sugar water with protein powder and vitamins," Berger said, "and after a few hours it would ferment and we didn't want the birds poisoning themselves on bacteria-laden food."

A portion of the birds had to be sampled every day. They were weighed, and examined, and both ends—throats and rumps—were swabbed for virological and bacteriological samples. Toes were clipped for smears and a syringe was used to take a few fractions of an ounce of blood from the chickens and quail and canaries. This was to be done for every bird three times during the experiment, adding up to a target of nearly 600 samples.

Much of the lab analysis would wait until Sileo got back to Madison, but all the red blood cell tests and blood smears had to be reviewed immediately. This was Berger's job, and to do it she had to drive to a local medical clinic to borrow a centrifuge and then go to the University of Guam to borrow a microscope.

There were twenty protected birds and fifty-nine exposed at Potts Junction, thirty protected and forty-seven exposed at Ritidian, and thirty left protected in the DAWR lab. The enclosures were packed—"gang-caged" is the term Sileo used—and working in them in Guam's heat and humidity was oppressive. The protected cages, with their layers of hardware cloth, screening, and mosquito netting, were so dark, even at noon, that Sileo and Berger had to wear miners' head lamps when sampling and feeding the birds.

"You'd be in there, in shorts and a T-shirt with a head lamp," Berger said, "dripping sweat and you'd have to reach into cages and carefully handle these little birds."

Then as birds died—and they did—necropsies would have to be performed and tissue samples fixed.

"It was exhausting," Sileo said, "In fact, there was so much work involved in taking care of the white-eyes, canaries, and chickens, if we had had the full complement come in, I wonder if we could have taken care of them all anyway. We were stretched pretty thin those first couple of weeks."

Muna and another DAWR technician did help out, but most of the work was left to Sileo, Berger, and Savidge. "That was the most intensive study I had ever done," Savidge said. "We accumulated a horrendous number of samples."

But there was always time for a quarrel. Savidge now argued that, considering the reduced number of chickens and quail and the dexamethasone deaths, the lab birds should be moved to the field. Her concern was that if the sample size became too small, it would be impossible to achieve any statistical proof and the entire study would be useless. Sileo was reluctant to change the research design in the middle of the experiment. "You have a hard time explaining your methods when you change horse in midstream," he said.

"Another long, agonizing discussion with JS," Sileo wrote in his field notes. Later that day, he acquiesced, but in another compromise it was eventually decided to move just two birds from the lab.

Not far from the Ritidian site was a little beach of white sand where sailors from the navy's sub-listening post had strung a hammock. It was a spot where the sea breeze slid in off the reef and dozens of yellow butterflies played, and it was here Berger would retreat when the arguments or the heat in the cages got too intense.

In the midst of all this activity, the Fish and Wildlife Service announced approval of the DAWR's application to have ten bird species and two species of bat placed on the federal endangered species list. Of course, most of the birds had vanished from the island, and the white-eyes, flycatchers, and broadbills were in all likelihood extinct. Still, they were now on the list. "It took the federal government almost six years to the day to place them on the endangered list," Bob Anderson calculated. "Why did it take so long? I really don't know."

On August 16, fresh troops arrived for the sentinel study: a shipment of thirty chickens landed at the Agana airport, warm and clucking. They were taken to the DAWR lab and two days later were put into the field.

The day they were placed in the forest, two more canaries died. This time the cause of death was easy to establish. They were eaten by a snake. After feeding the birds, a DAWR staffer apparently failed to close one of the two latches that secured the door. That was enough for the brown tree snake. "It wedged itself in somehow," Sileo said. The canaries were on their suspended platform, but this was a five-foot snake. "It stood itself upright and got to the platform," Savidge said. By the time it was discovered, it had dined on two canaries. "Those things are amazing. . . . You can't do a disease study because the bloody snake eats your birds," Sileo exclaimed.

Birds—particularly those that had been injected with dexamethasone—continued to die, adding to the tension and confusion of the exercise. "Any time a bird died down there I got a natural reaction, not an overreaction," from Savidge and Berger, Sileo said. "First of all, they don't like it when things die. Secondly, they panic because they think our experiment is going to hell. So that was a constant source of frustration and argument."

Then in the third week, while examining the blood smears under the microscope, Berger saw something in one of the samples from an immunosupressed Saipan white-eye. It looked like a parasite. Perhaps there was a disease here after all. She showed it to Sileo. "Great Scott," he thought, "there is a disease on Guam."

"I wasn't sure what it was, but I was afraid I knew what it was. I was afraid it was a plasmodium," he said. Plasmodia are microscopic, single-cell organisms, associated with a number of diseases, including malaria. These are "virulent pathogens and real trouble," Sileo said. It was, in fact, a plasmodial disease—avian malaria—that had been linked to the serious decline of birds in Hawaii.

Some smears were sent off to Ellis Greiner of the University of Florida at Gainesville. In the meantime, Berger went back and pulled the "reference" slides of white-eye blood Savidge had made the month before immediately upon catching the birds on Saipan. As she carefully wound her way through the maze of cells on the slide, embedded in a cell here and a cell there were the same parasites. They were not as numerous as in the blood of the dexamethasone-treated bird, but they were there. The parasite was not a Guam disease. It was a Saipan disease.

Within a couple of days, Greiner reported back. It was *Plasmodium circumflexum*, a malarial agent. Maybe they hadn't found a disease on Guam, but they had transported a new one to the island. "So here we are. Not only are our sentinels contaminated, but stand subject to the charge that we are introducing disease to Guam. That was probably the nadir of the whole project for me," Sileo said.

"What was curious was that the white-eyes were clearly dealing with this parasite. These were healthy birds. We probably would never have found it if the birds weren't immunosuppressed," Savidge said. "The plasmodium wasn't a problem for them."

But it was now imperative, Sileo thought, that Savidge not release the white-eyes on Cocos Island. He called his boss and laid out the problem. "Milt got concerned that we could be charged with releasing some kind of exotic disease. . . . He wanted those birds killed and when I told him I didn't think I had enough power to force her to do it, he started to go through Fish and Wildlife channels."

The irony was that once the parasite had been found Savidge had completely abandoned the idea of releasing the birds. "I felt terrible. Of course, there was never any question that those birds would have to be sacrificed," she said. But in an effort to avoid another fight, Sileo never broached the subject.

On September 6, Bob Anderson received a call from Bill

Kramer, the FWS Pacific island administrator in Honolulu. The message was simple: "Do not release the test white-eyes." Anderson passed the word to Savidge.

She was livid. "She felt I had gone behind her back. That outsiders were telling the DAWR what to do. I guess it was true. But under no circumstances was I going to let those white-eyes be released," Sileo said. In an attempt to escape conflict, he had used a back channel, and in doing so had provoked the very argument he had sought to avoid. The argument ended, Sileo remembers, with Savidge stalking off threatening to flush the white-eyes down a toilet.

The sentinel study finally trudged to its conclusion as all the surviving birds were brought in at the end of September. The tally ran as follows:

> Seventeen canaries died from infections and two were eaten by a snake. Fourteen of those canaries had been among the twenty-two injected with dexamethasone.
> Seventeen white-eyes died from infections. All seventeen had received dexamethasone.
> One quail and thirteen chickens died. Five of the chickens were killed in fights with other chickens.

The causes of death included pneumonia, anasarca, or dropsy, and a host of fungus infections, including infections of the heart and the liver. Five chickens came down with avian pox. But that was no surprise.

"We already knew that pox was on the island. But it seemed to be host specific, meaning that no other birds got it, just the chickens. It didn't cause all that much problem for them," Savidge said. None of the cases of avian pox among the chickens was fatal.

Two white-eyes did die from the plasmodium parasite, but they both had the disease in their system when they were captured on Saipan. "Apparently, their immune systems were coping with it until they were immunosuppressed," Savidge said.

There were equal rates of mortality among the protected and unprotected birds that received dexamethasone. "There was really no consistent cause of death so there was nothing that was suspicious," Savidge said.

Sileo was satisfied with the results. "The dexamethasone either killed them outright or so immunosuppressed them that they succumbed to normal avian fungi. They got all kinds of fungus infections, which confirmed that they were in fact extremely susceptible to infectious agents," he said. "But none of them came up with a 'mysterious Guam agent.' So I think they were extremely useful."

The study had its problems, particularly the parasite-riddled white-eyes. "The Saipan white-eyes are the best piece of information that we have that there was no Guam disease because they are so closely related to the endemic, native white-eye, and they were clearly immunosuppressed and susceptible and they didn't contract anything surprising," Sileo said. "On the other hand, they caused us so much concern because of the possibility that we introduced an exotic disease to Guam."

There was one more unpleasant exercise—disposing of the 127 living birds. Sileo was prepared to kill each and every one of them. While both Berger and Savidge understood the need for it, neither liked it.

"Wildlife biologists have no concept at all of histopathology," Sileo said. "Wildlife pathologists are a rare mix, because what makes you like to be a wildlife biologist makes you dislike sitting over a microscope all day. They had no concept of how much information you can extract from the microscopic examination of these experimental animals, but the only way you can get that data is to kill them."

Sileo killed the birds by exsanguinating them—sticking a needle into their hearts and sucking out all the blood. "We needed all the blood to get as much serum as possible for serological tests," he explained.

Berger strongly objected to the technique since it spun the animals into shock before death. "I realize that there is knowledge to be gained from histopathology," she said, "but we also have a responsibility to those animals to the end." She carried the birds to Sileo one by one, with tears in her eyes, sometimes saying good-bye to one or another by name. Savidge, no happier, occupied herself with other business.

On September 28, Sileo left Guam. He had missed his wedding anniversary but had managed to stop by the Micronesian Trading Company one of the island's gold brokers to buy his wife, Dottie, a present.

Along with the gold locket, Sileo returned to Madison with hundreds of samples—many of them frozen in liquid nitrogen. He was now ready to plow through the slides and vials and write the definitive paper on avian disease on Guam.

But he arrived back at the Wildlife Health Research Center to find another crisis brewing. The National Wildlife Federation had filed a suit to ban the use of lead shot in all waterfowl hunting in the United States because it endangered the country's bald eagle population.

The federation contended that bald eagles were eating ducks and geese that were carrying lead pellets, and the eagles were getting lead poisoning. The eagle was not only the symbol of the nation, it was an endangered species and thus protected by federal law.

The federation had based its case on data amassed over the years from necropsies done at the Wildlife Health Research Center on dead eagles. The environmental group obtained the data through a federal Freedom of Information Act request. Sileo remembered federation people busily Xeroxing up reams of files at the center just before he left for Guam.

One of the biggest constituencies of the Fish and Wildlife Service is hunters, particularly waterfowl hunters. A huge chunk of service funds comes from the purchase of "duck stamps," which must be affixed to waterfowl hunting permits. The end of waterfowl hunting was no joke for the service.

"So, I come back and there is Milt sitting on the biggest crisis of his career," Sileo said. Now, it turns out that the man who had done all the eagle necropsies for more than a decade was Louis Sileo. So once again, crisis dictated the agenda. Sileo was told to put aside the Guam work and analyze all the diagnostic work the center had done on eagles for the last twenty years. "I was really pissed," he said.

The Guam specimens were dispatched to big stainless steel refrigerators. Some of the tests weren't done for another two years. Some of the serum that those little birds gave their lives for was lost. "It turned out that it wasn't needed because there simply was nothing in the samples we did test," Sileo said

Both Savidge and Sileo believed the work was statistically strong enough and scientifically complete enough to address the disease hypothcsis. And despite the scientific brush war, or perhaps because of it, the two had become good friends.

And years later when all the tests were done, all the data analyzed, and the report finally written, Sileo and Savidge concluded that there was no evidence of a mysterious Guam disease.

11

November 1984

F all had just about left the Blue Ridge Mountains. The foliage had peaked, and much of it was already gone, leaving only a gauzy covering of bare branches on the hills. It was in this sylvan setting, with its presage of the kind of winter Guam never knows, that Beck, Shelton, and a small band of biologists met to ponder the future of kingfishers and rails.

Beck had just arrived on the mainland with the second shipment of birds—nine kingfishers and twelve rails. The group was already well on the way to establishing its target population of at least thirty kingfishers, and while capturing rails had proven more difficult than expected, the birds at the DAWR facility had already begun to mate.

What was next? That was the topic on the agenda for this meeting at the National Zoo's Conservation and Research Center in Front Royal, Virginia. Once a U.S. Army cavalry remount station and then a U.S. Department of Agriculture beef research facility, the 3,150-acre spread became the territory of the National Zoo in 1975. The USDA cattle were carted off and now Bactrian camels, reindeer, zebras, and South American emus graze on the rural Virginia hillsides.

Here, seventy-five miles from the main zoo, the business of analyzing the biology and husbandry of threatened species from around the world goes on. This is the home of rare Arabian oryxes, tree kangaroos, and Manchurian cranes.

Researchers at the center ponder questions such as the survival of the alpine ibex and mate choice in pintail ducks. They do this without the gaggles of gawking school children, the throngs of tourists, and the considerations of mounting crowd-pleasing exhibits that are part and parcel of zoos. And that's just the way Scott Derrickson likes it. Derrickson, a lean, chain-smoking figure with wire-rimmed glasses, graying temples, and an all-business demeanor, had succeeded Guy Greenwell as the zoo's curator of birds. He was passionate about biology and largely indifferent to the public relations aspects of zoos.

Derrickson had come to the zoo from the Fish and Wildlife Service's research center in Patuxent, Maryland, where he had been working on a project to reintroduce captive-bred, endangered Mississippi sandhill cranes into the wild. He was the host for this meeting and about to become a key figure in deciding the fate of the Guam birds.

The group gathered in a rustic lodge to ponder the future of the growing populations of captive Guam birds. How would the birds be managed? How would additional zoos be brought into the project? It was decided that the first step was to establish two separate programs—one for the rail and one for the kingfisher. Derrickson would quarterback the rail program, Shelton the kingfishers. Each man would be responsible for keeping detailed records on the fate and progeny of every bird.

Stud books were begun immediately to track the mating and offspring of the birds and it was also decided to submit petitions to the American Association of Zoological Parks and Aquariums to create official species survival plans. From an ad hoc decision in the lobby of a North Hollywood hotel, the Guam project was about to become an institution.

Gene Morton, the National Zoo biologist who had gone to Guam to help Beck trap rails, argued strongly that plans had to be made as soon as possible to find some way to return the birds to the wild. "I pointed out to Beck that there was no point in breeding these animals in captivity unless there was a way of getting them back into their native habitat," he said.

Beck had been running flat out to capture and care for rails and kingfishers. Savidge was now completing her snake study and preparing to write her dissertation on how the brown tree snake was wiping out Guam's birds. But Morton was right: someone had to think about a reintroduction program or else the Guam birds would be stranded in American zoos.

The group made one other important decision—all the birds brought to the mainland and their progeny, in perpetuity, would be the property of the island of Guam. There were two reasons for this position. First, it was the stated goal of the project to captive breed the birds only until the snake problem could be solved or other habitats could be found. Second, it would make the management of the birds in the United States easier. Sometimes zoos develop sticky fingers and are unwilling to part with popular property. "A zoo pays $80,000 for an animal or makes a major investment in a display and then is asked to send the animal to another zoo for breeding under a species survival plan and you can have problems," Derrickson said. There would be no such problems with the rails and kingfishers.

There would, however, be lots of other problems. Even though there were already twenty rails in captivity, and breeding seemed promising, the chances of catching many more appeared slim. In addition, a number of the birds were related. It turned out that there were no more than eleven independent breeding blood lines. The rails were facing a "genetic bottleneck" and the prospect of unavoidable inbreeding. It was the very kind of exercise that researchers like Nate Flesness, John Ballou, and Katherine Ralls had warned against. Special care would have to be taken with the birds.

As for the Micronesian kingfishers, numbers weren't the problem. There were already twenty-one birds on the mainland and every prospect of catching another dozen on Guam. In fact, Beck would return in January 1986 with another eight birds. The birds were being captured in three geographically distinct areas, which assured some variation. In addition, some birds had actually been taken out of the wild as mating pairs.

Within two years, the DAWR had trapped and transported to the continental United States a kingfisher population genetically diverse enough to develop an efficient captive breeding program. No, the problem wasn't numbers. The problem was just about everything else.

• • •

The forest birds on the Pacific island of Guam mysteriously began to disappear, until a Micronesian kingfisher (above) and the Guam rail (below) were the last native species in the rain forests.

The hidden culprit in this avian massacre was the brown tree snake, a native of Melanesia, which had been accidentally transported to Guam. In its new home, ample prey and no natural predators enabled the snake population to explode.

A denizen of the night, the brown tree snake slithered through the trees, swallowing lizards, bats, birds, and, as seen here, birds' eggs.

5

Julie Savidge, a University of Illinois graduate student in biology, was hired by Guam's Division of Aquatic and Wildlife Resources (DAWR) to solve the riddle of the island's vanishing birds.

Wildlife biologist Bob Beck was given the task of rescuing the few surviving rails and kingfishers from the forest and coordinating the development of a captive breeding program for both species.

6

7

Many biologists were convinced that a disease, not a snake, was responsible for the decline of Guam's birds. Julie Savidge collected blood samples from birds, such as these black francolins, in an effort to find a bacterium, parasite, or virus.

8

In a grand assault on the disease theory, Savidge, with the aid of the federal Fish and Wildlife Service, launched a massive "sentinel study." More than 100 birds, housed in special cages in the forest, were left exposed to possible disease agents. None caught the alleged "Guam disease."

Much of the research and trapping of birds and snakes took place in the thick rain forest beneath the cliffs of Ritidian Point—the last bastion of the birds of Guam.

News of the brown tree snakes' invasion brought federal Fish and Wildlife Service biologist Tom Fritts to the island. Fritts suspected that the tree snakes could spread to other islands, and finding snakes near airplane cargo containers confirmed his suspicion.

The snakes' ability to climb power cables also drew Fritts's interest. Snakes hitting electric power lines caused nearly 1,000 power outages on the island between 1978 and 1990.

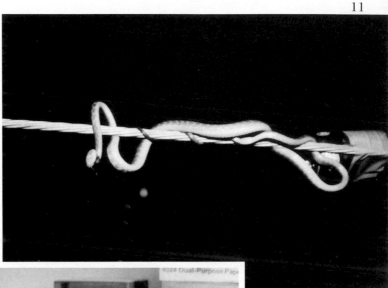

In an effort to learn more about the snakes, DAWR biologist Gary Wiles, shown here with veterinarian Diana Berger, implanted transmitters in some animals and tracked them by radio. He found that the snakes could live in just about any habitat on the island— even underground.

12

13

The captive breeding program for the Guam rail was so successful that plans were made to release some rails on the nearby island of Rota.

To help design and fund the rail release, Beck (right) enlisted the help of University of Tennessee ecologist Stuart Pimm (left). Pimm's graduate student Greg Witteman (center) was dispatched to Rota to oversee the release. Behind them is the project's infamous truck, which was driven 8,200 miles by persons unknown before a single rail was released.

14

Greg Witteman, shown here with DAWR's Herman Muna, planned to use radio tracking equipment to follow the progress of the birds. But technical failures and high mortality rates hindered the experiment.

Some of the first rails were released by S. Dillon Ripley, the former secretary of the Smithsonian Institution and a world-renowned expert on rails. Ripley's initial support had helped get the Guam captive breeding program going. He is flanked by Herman Muna (right) and Stan Taisican (left), a wildlife technician on Rota for the Commonwealth of the Northern Marianas Islands.

When the first batch of kingfishers had arrived in Philadelphia the previous January, they were taken to the annex of the Penrose Laboratory—a stone building that houses the zoo's hospital and a laundry. The birds were placed in a sliver of a room—twenty-five feet long and eight feet wide—with beige cinder-block walls and one-way windows for observation.

"That night we put in food," said Shelton. "Some of the birds immediately started eating, we saw through the one-way glass. So, things looked pretty good. We had no ages on them, because once they are in their adult plumage, they are in their adult plumage."

For the next two months, the birds were kept in isolation in the lab annex. Shelton organized a battery of volunteer docents to watch the birds. "We monitored everything," Shelton said, "how many mealworms, how many grubs they ate, weights and blood samples, everything."

As the weeks passed, all the signs looked encouraging. The kingfishers were eating and there were no signs of disease or stress. But that didn't allay Shelton's concern. "We knew we were dealing with a difficult species that had never been bred in zoos before," he said.

In fact, except for the kookaburra, the Australian "laughing jackass," whose high-pitched rolling cackle seems to be the required audio accompaniment to all movie sunrise scenes in the Australian outback, there was little experience in breeding birds in the kingfisher family. The Brookfield Zoo in Chicago had raised a grayheaded kingfisher and the Denver, San Antonio, and Bronx zoos had all taken a stab at breeding African pygmy kingfishers.

There was one other place that the pygmy kingfisher had been successfully raised—in the Greenwich Village apartment where Larry Shelton used to live.

Shelton had started his working life as a public relations man, not a zoo curator. But from his days as an Indiana farm boy he had been enamored with birds. "My family thought I was crazy," he said. As a teenager he wrote to James Peters, a noted ornithologist and professor at Harvard University, asking advice about a career in the field.

Peters wrote back advising him to find another line of work, something with more opportunity. There was, he said, little chance of making a living off a love of birds. "My family said, 'See? We told you,' " Shelton remembers.

He majored in liberal arts in college and then went to New York City, where he built up his own public relations firm. But the fondness for birds never left Shelton, and in his home he took on the challenge of trying to rear and maintain the most exotic species. At times he had as many as 200 birds in his rambling apartment. Two large planted aviaries, flooded with the glow of pale purple grow lights, dominated his living room. Each held up to seventy-five birds. Shelton was the Bird Man of Greenwich Village.

He raised tanagers, honeycreepers, thrushes, lilac-breasted rollers, and, of course, pygmy kingfishers. "They were the kinds of birds that zoos wanted," he said. Once he had mastered the breeding and husbandry, Shelton would donate the birds to zoos and go on to some new species, some new challenge. He donated several birds to the Bronx Zoo and became friendly with the curators there.

In 1976, Shelton sold his business and decided, at the age of forty-five, to take an early retirement in the California sun. He began by "drastically reducing" his collection of birds. The availability of so many of Shelton's flock piqued the curiosity of the then curator at the Bronx Zoo, Joseph Bell. When Bell learned that Shelton had sold his PR firm, he suggested that he apply for the recently vacated post of curator of birds at the Philadelphia Zoo.

"I did it as a lark," Shelton said. "I never expected to get it. The moving men were in my apartment, packing up for San Diego, when I got the call that the job was mine." Now, Shelton would have to apply all the skill he had gathered in his Greenwich Village apartment.

There are about eighty-five species of kingfishers spread around the world. They tend, like the Micronesian version, to have strong, compact bodies, short necks and wings, and pronounced beaks. It is believed that the bird originated in Southeast Asia, and it remains primarily a tropical bird. Members of the family can be found in Africa, the Pacific, and the Americas. Two species—the belted kingfisher and the green kingfisher—live in the continental United States.

The ancient Greeks believed that the birds nested at sea and called them *halcyon*, which means "the one that conceives at sea." The cinnamon-colored Micronesian kingfisher's scientific name is *Halcyon cinnamomina cinnamomina*. Reality is, however, more mundane than myth: the kingfishers actually nest in sand banks, termite hills, and trees.

There are water kingfishers and forest kingfishers. The birds got their names because fish are a major part of their diets. But here the Micronesian kingfisher parts company with its cousins. It doesn't eat fish at all. How else did it differ? No one really knew.

"We tried to learn everything we could about their natural ecology. We knew they excavated on the wing and so needed soft material. . . . They don't have pneumatic padding. They'd get a Tylenol headache trying to pound away on hard wood," Shelton said. "But there were so few left in the wild to study that there was a lot we didn't know."

According to past field observations on Guam, a kingfisher pair excavate their nest together. The birds appear to mate year-round except for the summer rainy season, producing a clutch of two eggs. The goal of the zoo curators was to evoke the same behavior and fledge as many birds as possible. The question was how to do that.

After their quarantine, Philadelphia's first pair—bird Number 1 and bird Number 2—were moved to a cage in the bird house thickly planted with bamboo, ficus, and palms. Shelton went down to neighboring Fairmount Park and picked out a dead log, already stripped of its bark, and put it into the cage.

The birds flourished on a diet of lizards, crickets, and worms. Shelton even tried minnows to see if the kingfishers would eat fish. They did. But they didn't display any interest in making more kingfishers. Shelton wasn't alarmed. "We didn't expect nesting the first year," he said. The birds at the Bronx Zoo showed no more enthusiasm for procreation than did their Philadelphia brethren. That was the state of affairs through the spring and summer and into the fall, when Bob Beck arrived with the next shipment of birds.

"I was walking around Larry's facility, talking about why the birds hadn't bred," Beck said. "I took one look at the logs he had for the kingfishers and I realized that they weren't soft enough."

"It was," Shelton said, "a question of how soft is soft." Shelton went down to Saint Catherine's Island off the South Carolina coast and came back with some near-compost palmetto logs. Gene Morton brought back logs for the birds at Front Royal, Virginia, from a trip to the Louisiana bayous. The Bronx Zoo found similarly decaying boughs.

Just four months later, in March 1985, there was a nest at the Philadelphia Zoo. The eggs proved to be infertile. "We were worried they were too old to reproduce," Shelton said. But in May, the Bronx Zoo successfully hatched chicks. Within a few weeks the Philadelphia Zoo had done so too and by the fall so had Front Royal. It had been more than a year since the birds had arrived on the mainland, but now it looked as though the zoos could manage to breed them.

Then something really strange began to happen. The Bronx Zoo hatched its first two chicks at the very end of May. The next day when a keeper looked into the nest one chick was gone. The following day the second chick had also disappeared. It looked as if they had been eaten by the parents. The same thing happened to two more chicks in July.

Chicks born that same summer in Philadelphia were reared by their parents with no problem, but in September, the same sort of chick cannibalism happened at Front Royal. Why were the birds eating their young?

"It probably has to do with some stress in the captive breeding situation. It is hard to imagine that kingfishers actually cannibalize their young in the wild," Beck said. "It seems very unlikely, although we don't have the data from the field to say one way or another."

Shelton, while troubled, wasn't totally surprised. "Cannibalism happens to a lot of carnivorous birds in captivity when they are stressed," he said. Rollers, for example, are related to kingfishers and have been captive bred a number of times, with several incidents of cannibalism. "What can trigger it is not apparent to us, maybe not enough food at the right time, a noise in the night," he said.

One suspect in the kingfisher case has been the pinkie, a small, hairless mouse that many of the zoos had begun to use as a staple food for the birds. Pinkies look a lot like chicks. Perhaps, some speculate, after a steady diet of mice, the birds were mistaking their offspring for just one more pinkie.

"Some zoos have cut pinkies out," said Beth Bahner, the animal collections manager at the Philadelphia Zoo and the kingfisher's stud-book keeper. The best food is the anole, a small lizard. But they are far more expensive than pinkies. The mice can be mass produced, but the supply companies have to pay kids to go out and catch the anoles. Hence the cost, and also the uncertain supply.

"Trying to get a steady supply of anoles in the winter is really hard," Shelton said. There had been some discussion of trying to put the birds on nonliving food. "But," he argued, "my feeling is not to try to get them on to nonliving food. We would be breeding out of them some of their natural behavior."

Some zoos, such as the Bronx Zoo, opted to hand-rear the birds, removing them from the nest just before or just after hatching. But like trying to change their diets, hand-rearing, some fear, poses a risk in changing the birds' basic behavior.

"One problem," Beck said, "has been the propensity of some zoos to want to hand-rear birds. Shelton never wanted to hand-rear his birds. But some of these zoo folks wanted to rear all the birds by hand. It is the kind of thing these zoo folks do. They weren't initially oriented to this conservation approach of trying to get the birds back into the wild."

Nevertheless, as the cannibalism continued, hand-rearing became an increasingly popular defense. Between June 1986 and June 1991, thirty-eight chicks were killed by their parents. And it happened at all kinds of zoos, as the breeding program expanded. Chicks were lost in St. Louis, San Antonio, Cincinnati, Chicago, Washington, D.C., and New Orleans. Even the Philadelphia Zoo, which had initial success with parent-rearing, lost some birds to cannibalism.

"The point is nobody knows what the stress factors are that are doing this," Beck said. "Some zoos have had more success than others. But there seems to be no correlation between whether the birds are on display or not on display, whether they have been hand-reared or not hand-reared the generation before. It is hard to figure out what is going on here."

Some kingfishers have been admirable parents and reliable contributors to the gene pool. For example, there is the performance of kingfisher Number 28.

Number 28 was caught by Beck and Muna in late 1985 at Northwest Field and brought to the Philadelphia Zoo in the last shipment of kingfishers on January 31, 1986.

When he arrived, he weighed fifty-four grams—just under two ounces. He was paired with kingfisher Number 29, who had been caught at the conventional weapons range at Andersen Air Force Base. In March, they had one male chick, Number 68. The following September, they had a second male chick, Number 88. In both cases, Number 28 and Number 29 diligently raised their young.

When they were old enough, Number 28's offspring were sent to other zoos. Number 68 was packed off to San Diego and Number 88 was sent to Houston.

By the time those birds were hatched, the kingfisher program was steadily expanding. First, Denver was added. Then Saint Louis. Then more and more zoos joined—San Antonio, Kansas City, Milwaukee, Cincinnati, Chicago's Brookfield and Lincoln Park, Tampa's Lowry Park, the Pittsburgh Aviary, New Orleans's Audubon Park, Orlando's Sea World, and the Riverbanks Zoological Park in Columbia, South Carolina. By 1991, seventeen institutions were working with the species.

At first, Shelton was very careful in parceling out kingfishers. "We took excessive precautions," he said. They didn't even fly on commercial flights. The first few shipments, like the one to Denver, flew on an ARA Services Inc. corporate Lear jet, thanks to the largess of the Philadelphia-based food services company.

In 1988, Number 29 died of oophoritis, an inflammation of the ovaries. Number 28 was then paired with Number 61, who had been born at the Bronx Zoo in 1986. Her parents had been trapped by Beck in the Ritidian forest.

By this time Number 28 and Number 61 were living in a special greenhouse that had been constructed behind the bird-house expressly for the Guam bird project. The money for the building came out of the checkbook of a wealthy and anony-mous zoo patron who heard the story of the Guam birds from Shelton at a cocktail party.

The greenhouse—a white, prefabricated, paneled structure about forty feet long—sits unobtrusively behind the birdhouse. Except for the sign proclaiming it "The Guam Bird Propagation Center—Not Open to the Public," it might easily be mistaken for an equipment shed.

Inside are eight large, floor-to-ceiling cages built of wooden planks and wire mesh. There are some potted plants on the ce-ment floor and a large plastic basin of water for drinking and bathing. In the upper corners of the cages are logs about 20 inches in diameter. They are soft, soft enough to easily push a key into. That is now the zoo's softness test. Banks of sun lamps are ranged on the ceiling—a backup heat source—and the con-stant hum of an exhaust fan provides background noise. All very utilitarian, but it does the job.

It was here in cage 5 that Number 28 made his home. On May 10, 1989, Number 28 and Number 61 hatched a chick. Four days later it was gone. Cannibalism. In late June, the pair had another clutch. But before the incubation was completed, both eggs were found on the floor of the cage. Both were so badly damaged that they could not be saved.

Things weren't working out with Number 61. By early 1990, Number 28's weight was up to eighty-five grams—nearly three ounces. The sign of an unhappy relationship? Perhaps with hu-mans, but the Philadelphia Zoo keepers insisted it was just too many wax worms, which are notoriously fatty, sort of avian junk food. Number 28 went on a diet and got a new mate.

Number 38 had issued from two of the original birds Beck brought to the mainland from the Ritidian Forest. Hatched in Front Royal, she had spent two years at the Pittsburgh Aviary with Number 47, a bird hatched at the Bronx Zoo.

It took four months before Number 28 and Number 38 started nest building. Their first clutch—a single egg—was infertile. But the second clutch yielded a chick. It had scissor bill—a defor-mity in which the top and bottom bills do not line up. The cause might have been congenital or nutritional. The bird was reared by the parents and the problem largely corrected itself.

Number 28 had added another new number to the kingfisher reservoir. Indeed, by that time Number 28 was a grandfather, as Number 66 had sired five chicks and Number 88 had eleven. In all, Number 28 had produced twenty-one descendants, eleven of whom had survived.

The zoo keepers had enough confidence in Number 28 to let him raise his young. But when Number 61 and Number 111 had a fertile egg a few months later, the zoo kept a close eye on it. Twice a day a keeper would climb a ladder to check the nest. In the morning, he would also wash down the cages and bring a meal of mice, crickets, and worms.

It was decided to pull Number 111's egg and incubate it. Three days before it was expected to hatch, a keeper climbed up to the nest, took the egg, and brought it to the incubation room on the second floor of the birdhouse. The egg was placed in a "hatcher," along with other eggs. When the chick emerged, its eyes and ears were sealed and it could not regulate its own body temperature. It was almost as if it were cold-blooded.

It was plunked into a hospital cage, which looks like a portable television and keeps the bird at a temperature of 99 degrees Fahrenheit. The naked little chick—weighing no more than a quarter of an ounce—was placed in a small bowl with a terry cloth liner. It was hand-fed small pieces of chopped pinkies eight times a day. Then a few days later bits of mealworms, wax worms, and crickets were added. Slowly the number of feedings were reduced to only two after a little more than a month.

Day by day, the temperature was reduced by one degree, until it reached 85 degrees. By that time, the young bird could control its own body temperature.

Notwithstanding the best efforts of Number 28 and the zoo curators, raising kingfishers has been a struggle. The problem of cannibalism has been compounded by other deaths. Almost immediately three of the original founder population died, reducing the core group to twenty-six.

By 1993, twenty more founders would die from a variety of diseases and traumas. In all, sixty-seven birds would die from an assortment of complaints, including pneumonia, septicemia, liver disease, and choking. But the most menacing development was an outbreak of avian tuberculosis. It first surfaced at the Saint Louis Zoo, and Shelton immediately quarantined the institution, allowing no more birds in or out. But there were also outbreaks in Cincinnati, Front Royal, and Philadelphia. By 1992, there had been seven reported cases of the highly contagious disease.

The emergence of this tuberculosis created a dilemma for the breeding program. The only way the disease could be transmitted was from bird to bird. But there is no test for the disease or way of diagnosing it before the bird is in its grip. Taking birds suspected of being infected out of the program would ultimately bring the entire breeding effort to a halt. But pairing them might spread the disease around.

"If there were more birds we might isolate or euthanize birds we thought might be infected," Bahner said, "but the population is so small, we can't afford to ban birds. We can't afford to throw them away."

As a result of the limited pool of birds, an individual suspected of being exposed to TB is isolated for six months and screened twice, and then if no symptoms emerge it is sent on its way back into the population and from zoo to zoo.

Because of the cannibalism and disease, the kingfisher population has grown fitfully. From an original population of twenty-nine birds trapped on Guam, the population rose to sixty-two in 1988. But in 1991 it had increased by only three to sixty-five. By 1992, the population had fallen to fifty-six and prospects for the future were uncertain.

"The bottom line is that we've not been able to breed the number of birds we've wanted, and some lines have not bred at all while some lines have bred a lot," Beck said.

Reviewing the status of the birds in late 1993, Bahner said, "Truly, this is now a race to prevent extinction."

12

June 1985

om Fritts was back, just as he knew he would be, and this time he came to Guam loaded for snake. He had a multitude of questions he hoped to answer. He wanted to know how many snakes were out there. How they lived. How they could be trapped. But most of all, Fritts wanted to get inside the mind and heart of the animal. "I wanted to know what drove the snake. I wanted to find the chink in its biological armor," he said.

Fritts's goal was to develop some kind of snake control program, probably a way to trap the animal. He also came with 500 sticky-board rodent traps, which he distributed to bewildered customs and military transport officials, suggesting they spread them around warehouses and cargo bays to guard against the snake.

But despite Fritts's best efforts to convince these officials of the seriousness of the snake problem, they remained skeptical and the traps remained in their boxes. Fritts might consider the snake a menace to the entire Pacific, but no one else did. Yet.

Since finishing the sentinel study the previous fall, Savidge had devoted herself to the brown tree snake. She had been collecting and collating data on all the snakes she caught in her quail traps and those that were brought to the DAWR by island residents looking for "the snake lady." In January, she had created a grid of thirty traps, spaced about eighty feet apart, in a forest patch near the Navy Communications Station. After eighteen nights of trapping, Savidge estimated that there were about 6.5 snakes per acre.

A thirty-trap grid was a start, but Fritts was planning to "trap in a big way." Thirty would be a starting point. But he would go to fifty and then eighty traps, and eventually one hundred, because he was thinking of ultimately trapping in a really, really, really big way.

Savidge had pioneered brown snake trap technology. Could Fritts improve upon it? He brought two designs to Guam. One was a tube of wire screen, much like Savidge's. It was about thirty-two inches long and twenty-two inches in diameter, with funnels at each end. The other trap, an attempt to use some military surplus, was a hexagonal cylinder of plastic mesh with a cap at one end and a hardware cloth funnel at the other.

Fritts wanted to see which trap worked more effectively. The big question, however, was what to put into it. Savidge had used quail to demonstrate rates of predation by the snake. But if the goal was merely to lure the snakes into the traps, quail were less than ideal bait. "They tended to attract other things, like monitors," Fritts said. "You need to give them birdseed and water and protect them from heavy rainstorms and hot temperatures. They were like problem children. It wasn't feasible to have a big trapping program with such a scarce and fragile bait.

"So we began asking, 'Is there something artificial we can use to bring them in?' " Fritts explained. The first bait he turned to was "chicken litter," the droppings and wood chips at the bottom of a chicken coop.

At the core of Tom Fritts's very practical problem was a very fundamental question: What was the sensory makeup of the tree snake? What turned it on?

"We tend to classify snakes by whether they are visual or olfactory. Racers are very visual, and so are cobras," Fritts said, "whereas rat snakes and corn snakes are by and large not very visually sensitive."

It was hard to figure out what the brown tree snake was. On one hand, in captive settings it appeared visually alert. "It was definitely nervous and would follow you," Fritts said.

Yet it was taking birds out of the trees in the thick blackness of rain forest nights. One of the ways visual snakes hunt is by ambush, but birds sleep at night, so the ambush strategy didn't make much sense. There had to be some other cue the snake was using to find birds. Fritts wanted to find that signal.

Outside of the snakes trapped, accidentally encountered by islanders, or pulled from cliffs by Muna, the biologists still didn't know much about the "gob" of brown tree snakes living on Guam. For example, while they suspected that brown tree snakes might have two clutches a year of six to twelve eggs, no one knew for sure.

In an effort to fill in at least some of the blanks, the reluctant Gary Wiles took to the field again to radio track snakes. This time he used a fingernail-sized transmitter without a wire antenna. "I tracked three snakes and this time I got good results," he said. "Each one of those snakes behaved differently."

The first snake—a little one, slightly more than four feet long—was released up at Northwest Field and Wiles tracked it for about a month. "It stayed in a very small area. Each day it would move to a new resting site—within twenty-five to seventy-five meters," he said. But what was really interesting was the sites that it chose.

He found that the snake used a wide variety of resting places, "from the very tops of cycad trees to the clawlike roots of pandanus trees."

One day, Wiles said, he tracked it "to a small leaning tree that had a big clump of ferns that had formed a big root mass. The snake had gone into the root mass and was sleeping in there. It had a whole variety of sleeping sites, all of which were dark and cool."

Perhaps this did nothing more than confirm what Savidge and Fritts had already known or suspected. But with each of the next two snakes, Wiles gleaned bits of new information.

The second snake was also released at Northwest Field, where it had originally been captured. It immediately went underground and stayed there. Wiles described the snake this way:

This guy had found an old dump site, where the military had dumped a whole bunch of cans. They looked like paint cans. These had eventually worked their way into the soil, and it was kind of a buried dump. So the snake had gone underground and was sleeping in one of these cans. He was under there for about ten days.

Every day I went up there, and he was in the exact same location. I didn't know if he had died or what was going on, so I dug him out and he popped out and slithered away. He was inactive for some reason. The next day he was fifty yards away or so, underground, sleeping in the cans again, and he stayed there.

I did come out on a few nights and found that he was going and coming and feeding. But he always went back underground.

What Wiles found interesting about this snake's behavior was that it showed that there was no simple way to secure even a piece of land from the animal. "You could clear-cut an area of forest to get rid of snakes, and they could still be hiding under the ground," he said. "So just clearing the vegetation doesn't mean you are going to get rid of the snake."

The third snake was one captured by Wiles and Beck in a cave down at the Naval Magazine. "We wanted to find out how snakes behave in caves because we have this swiftlet population in a cave down there," Wiles said. "It just happens Bob and I found a snake in another cave. It was a nice big cave, and he was deep inside."

After bugging the snake, Wiles returned it to the cave. The next day it was gone. Using his radio tracking equipment, Wiles searched everywhere and finally found it on the hill at the back side of the cave. The snake must have slithered out through some crevice in the cavern.

The animal then used small cliff faces on the hill or the base of boulders as resting spots. Just as it started to move, the radio began to give out. But Wiles found it in a patch of sword grass. This was another revelation. The birds that had managed to survive on Guam—such as yellow bitterns and blue-breasted quail—were field dwellers, and it had been generally assumed that the savannas were a less hospitable habitat for the snake.

The sword grass was about six feet tall, and at its base was a layer of dead grass. The snake was under that layer. "It was a hot day," Wiles said. "I took the temperatures. . . . In the grass it was

105 degrees, but down under the dead vegetation, where he was, it was 20 degrees cooler."

Despite the promising results, Wiles abandoned the radio-tracking work. In part, he was concerned that in some way his data might be biased. In part, he was more interested in following his bats, whose population once again appeared to be on the wane. By the summer of 1987, when Wiles completed the second round of radio-tracking exercises, the bat population had fallen to somewhere between 500 and 600 animals.

The bats were still suffering from illegal hunting and the occasional typhoon, but Wiles had found a new menace and one that explained the absence of juvenile bats from the colony—the brown tree snake.

Bats deliver their young while hanging upside down. Wiles had patiently watched bat births through a huge spy scope. "It seems like it takes twenty or thirty minutes before the whole baby emerges," he said.

"First comes the head. Then it comes out with the wrist of one of the wings and then pretty soon the whole wing is out and then the other one and then the rest slides right on out and it crawls down to the mother's face or one of the nipples. Bats actually have nipples under their armpits rather than on the breasts and then the baby latches on right away and gets carried like that."

Even when the mother flies off into the night in search of wild figs and breadfruit, the baby goes along for the ride. "The problem apparently comes when . . . the young become too large to be carried around all night," Wiles said.

"As the young bats get bigger, the mothers might fly for a bit, park their offspring in a treetop, hunt for fruit, and come back and pick them up," Wiles said.

"I was once hiking through the woods and I had just been at the colony. I had camped out overnight and I heard this strange calling down there. . . . Then I heard the calling a couple of days later out in the middle of woods away from any known colony—this was out on Northwest Field—and I walked to it. There was this baby bat, it was about eleven o'clock in the morning, and he was crying his little heart out. He was probably very hungry and very afraid. The baby was left behind; maybe the mother was hunted or for some reason she never got back." Wiles took the baby back to the DAWR compound and hand-reared it. The bat still lives in a cage in the compound.

When the young bats got a little bigger they were simply left in the colony, swinging from branches for the entire night. These little bats, Wiles was certain, were easy prey for the brown tree snake. That is why, he believed, he could see newborns and adults in his scope but never any juveniles.

One thing was certain. There were snakes in the area. The bats made their colony in virgin forest that rimmed the northern tip of Guam. To watch them, Wiles could climb about a third of the way down the rugged limestone cliffs, perch in a rocky cranny, and peer through his large scope at the treetops below. Once while intently studying the colony, he felt something brush his pants leg. He absentmindedly swiped at it. He felt it again. He looked down to see a brown tree snake crawling up the inside of his trouser leg.

Wiles knew that tree snakes could be anyplace. The radio-tracking data had shown that. "What we found was that they were using a wide variety of habitats, all over the island. They were everywhere." But it wasn't easy to find them. They rarely crawled up one's pants. Even with the help of a radio transmitter there were several times, Wiles said, when it was difficult to visually locate the animal.

But they were out there. Everywhere.

By the time Tom Fritts returned to Guam, Julie Savidge was in the home stretch of her dissertation. In December 1984, she and Sileo had presented their tentative findings to a Fish and Wildlife Service conference in Honolulu, and these had been favorably received.

She was now amassing data on the brown tree snake at a blazing clip. What she had gathered was very basic. But considering nobody knew very much about the animal in the first place, that was fine.

One simple thing Savidge did was measure the length of 125 snakes she had caught in the forests at Ritidian, Andersen Air

Force Base, and Northwest Field. From the snout to the anal vent, near the tail, they ranged from twenty-one inches to six feet. The majority were between three and three and a half feet. But there had been snakes reportedly caught in urban areas that were eight to ten feet long and weighed more than five pounds.

After performing necropsies on more than 200 animals, she had found that the snake's food was about one-third birds and bird eggs, about half was lizards and lizard eggs, and the remainder was small mammals, such as rats, shrews, and rabbits.

But what was more interesting was that the snakes taken deep within the forest had no birds in them at all. In these areas, lizards—skinks, anoles, and geckos—made up nearly 90 percent of the snake's diet. In contrast, half the food found in the guts of snakes taken in and around farms and villages was birds or eggs. But not white-eyes, kingfishers, and rails. No, those were long gone. The snakes were eating poultry, chicken eggs, and caged birds.

The tree snake's appetite wasn't limited to this standard fare. "We opened one up and found three spare-ribs," she said. "One of the DAWR technicians camping down at Targue Beach one night was barbecuing chicken. She was talking and had turned away for a second and when she turned back a brown tree snake had come up to the grill and was trying to take a piece of chicken."

Together the size and the stomach content data had a disquieting implication. The small snakes were mainly eating lizards and living in the forests. But the snakes living in populated areas, where there was more food, were filling up on poultry, eggs, and small mammals and getting very big.

There was another troubling aspect that Savidge stumbled across. Island residents had reported a number of instances where snakes entered houses containing pet birds or entered pigeon and chicken coops shortly after eggs were laid. On two separate occasions, one pigeon breeder reported nine and fourteen snakes in his coop the day eggs were laid. Normally, he encountered two or three snakes in a month. Somehow the snake could detect the presence of fresh-laid eggs.

Ironically, this might have explained why Savidge's first predation experiment, her first summer on Guam, had failed. The quail eggs she used were store bought, infertile eggs. Perhaps they lacked whatever was the attractant that lured the snake. That, coupled with the very high rat populations on Rota, might have skewed the results.

In the spring of 1985, Savidge ran one of her original predation trap lines again. She was curious to see if predation rates had changed. Perhaps, having decimated its food base, the snake population had collapsed. She used her transect at the Navy Communications Station. In 1984, it had taken eleven days for 100 percent predation. In 1985, it took thirteen days. They were still out there.

While Savidge did her work, Fritts did his. With the help of Norman Scott, Jr., another FWS biologist, and a graduate student, Fritts ran a battery of experiments to test trap designs and the use of bird odors as a bait.

Using various combinations of his two trap designs with and without chicken litter, Fritts found the funnel trap to be the most efficient. In addition, the litter definitely improved trapping success. The project didn't end there. Curious about exactly how many snakes were out there, Fritts ran an intensive trapping experiment in July, with up to eighty traps on a single two-acre plot for seventeen days. He caught thirty-seven snakes.

Adding the snakes taken off that patch during the initial tests, Fritts counted fifty snakes taken out of two acres in forty-eight days.

He had done a fair amount of collecting in the Amazon, where as many as fifty different species of snake live together, but he had never encountered so many snakes as on Guam. "The results were fairly staggering," Fritts said. "I knew there were certain species of snake you only saw if you went out at night and indeed *Boiga irregularis* is pretty close to that. For every one you see during the daytime there are 100 out there at night."

Savidge had calculated a density of 6.5 snakes per acre. Fritts, with his more intensive trapping, raised the number to between twelve and twenty-five. In prime habitat, he figured that were probably 12,000 snakes per square mile—one of the greatest concentrations of snakes in the world.

Perhaps traps could be effective in guarding against snakes entering ports and storage areas, but given the scale of the problem, it was impossible to conceive of a large-scale trapping program to rid Guam of the brown tree snake.

"If you start out with two traps," Fritts said, "all you can do is build two more the next day and two more the next day and so on and your traps may be rotting and falling apart before you ever get to critical mass."

Just to clear snakes from 200 hundred acres, Fritts estimated, would take the equivalent of two people working full-time for two years. "What's happening during those two years is that the snakes are breeding faster than you can clear them," he said.

But Fritts wasn't, like some latter day Saint Patrick, looking to rid Guam of its snakes. He just sought some way to control the animal. "I think we started talking about what chemicals could be used. We knew that people had been talking about things like nicotine sulfate as a way to kill snakes," he said. "Nicotine sulfate used to be used as an insecticide, but it was banned by the EPA several years ago, because it is toxic to man and it can be absorbed through your skin. There is too much risk.

"Nicotine compounds are really good, because snakes are really vulnerable to nicotine, but so is man. I can absorb nicotine through my gums," Fritts said. "I am a snuff chewer so I know that. A tiny pinch of nicotine, a few grains out of a cigarette will kill a snake. But how do you present it to him in a manner that is environmentally safe?"

Pesticides might be useful in securing small areas against the snake but not in controlling it. "The biology was the key," Fritts said. "Was there something simple, something vulnerable about the tree snake?" Tom Fritts came to Guam for answers, but left with more questions.

13

November 1985

W hen the plane touched down in Denver, Bob Beck
had gained another 8,000 frequent flyer miles with Continental
Airlines. For the twelve rails traveling with him, it was strictly a
one-way proposition. They were sitting in cages in the now cus-
tomary seats at the back of the passenger cabin, during a layover, when the cleaning crew flung open the plane's rear door
and a blast of icy Rocky Mountain air came whistling in. Beck
jumped up shouting for the doors to be closed and then explained that tropical birds and skiing weather do not mix.

This was to be the last shipment of rails to the mainland. A total of twenty-one birds had been taken out of the forest on Andersen AFB as adults, juveniles, or eggs. The last rail was captured in March 1985. Trapping was abandoned in July. There
was one known rail left, but it lived in such a rugged area that it
was considered unreachable. Aside from that one bird, the
Guam rail appeared extinct in the wild.

There were, however, plenty of rails in captivity. Beck had
caught the first birds in 1983 and had built a compound next to
the DAWR headquarters—clearing an area about 400 feet by
400 feet of all brush and trees, placing a chain link fence
around it, and then covering the fence with a layer of hardware
cloth.

In the middle of this compound two aviaries—each about ten

feet by ten feet with ten individual cages—were constructed. The structures stood about three feet high and were covered with hardware cloth and netting to protect against snakes and mosquitoes. There was a large ribbon of open, close-cropped lawn between the fence and the aviaries.

By late 1984, nine rails had been captured and seven eggs taken from the wild. Soon twelve rail chicks were hatched at Beck's rail resort. Gene Morton brought four rails back to Front Royal, and by September they added another four chicks to the population.

Fruitful, fertile, and fecund—however you described them, the rails sure could produce rails. Lots of rails. Rails at a clip that made rabbits appear to have fertility problems. "We knew from the literature that these guys breed year round," Derrickson said. "But what we didn't know is what the breeding cycle of individual pairs is. Now we know they nest a couple of times in a row and then they take a break before they start to nest again. The female will start laying eggs within three weeks of hatching chicks."

At the end of 1984, there were a total of twenty-eight rails in captivity. A year later, there were sixty birds—a more than 100 percent increase—at the DAWR aviary, Front Royal, the Bronx Zoo, and the Philadelphia Zoo. By 1987, there were more than a hundred rails on Guam and at seven mainland zoos. Rails were breeding in Wichita and Baton Rouge and Louisville. The DAWR also had to build two new aviaries to hold another twenty-four birds.

Looking at the reproductive assembly line efficiency of the rails only underscored how absolutely voracious the brown tree snake had been.

"One of the reasons we wanted to have a whole bunch of different facilities breeding the rails and the kingfishers was that the curators would all have different ideas about breeding the birds and, if we had problems, somebody would figure out the right way to do it or it would be some combination of factors from all the different zoos," Beck said. "Well, with the rails it turned out that you could do almost anything and they'd do just fine."

• • •

There would have been even more rails if they hadn't spent so much time killing each other, for it turned out that they were as aggressive as they were sexually vigorous. There were repeated instances of adult pairs fighting, of adults going after juveniles, and of juveniles fighting among themselves.

"At the beginning we were losing two or three rails a year because of fighting," said Herman Muna, who among his many DAWR jobs cared for the birds.

Aggression plays a fundamental part in organizing the natural world. The competition among species for access to crucial resources and even among individuals of the same species for territory or breeding rights can lead to battles to the death.

In an article on the natural history of the Guam rail, Mark Jenkins speculated that "recorded instances of fighting, presumably among males, probably are related to territoriality."

But most of what Jenkins wrote, which was based on nearly twenty years of field observation by DAWR biologists, portrayed the rail as a good spouse, responsible parent, and upstanding citizen of the forest—not the crazed killer that the zoo curators were seeing.

The rail, Jenkins wrote, is a monogamous species in which male and female share the jobs of constructing their shallow nest of loose, interwoven grasses and roots and incubating their eggs.

The typical clutch consists of three to four eggs. A number of biologists had seen pairs of rails, their broods trailing behind, pecking their way through the woods.

Beck and Morton once saw an extended rail family trundling across a clearing. "We were lying in some shrubby area, where we had heard a rail. Being very, very quiet, we watched as a rail family group walked by us. Here came the two adults closely followed by their little black fluffy chicks and several immature birds that we could recognize because of their plumage. So it was a whole family group of several different cohorts of chicks and they were uttering this contact call, *gulp-gulp-gulp*, to keep the group together.

"That made perfect sense. These birds recycled every month," he said. "So you'd have chicks in the pen and they were one month old and your new eggs would be hatching on you and we saw this in the wild, too."

149

The bird's diet included snails, slugs, butterflies, insects, geckos, seeds, and palm leaves. "Adult rails may locate specific foraging spots and allow their chicks to peck there," Jenkins wrote, "often moving away to permit the juveniles to forage independently.

"Alternately, they may secure food (usually insects) and then allow the chicks to peck the items from their bills, or they may lay them in front of the chicks."

When they weren't feeding their young or themselves, building a nest, or hatching eggs, the birds spent the bulk of their time bathing and preening.

Rails, in general, are a highly successful, if little known, family of birds. They can be found on every major land mass outside the polar regions and many islands, where they have adapted to both forests and wetlands.

Of the 124 rail species, the Guam rail is on the large side, weighing about eight pounds and standing roughly ten inches tall. Still this was no hawk, no vulture, no terror of the rain forest.

And yet, in captivity, something had set them off. "They are very aggressive," Derrickson said. "One bird usually tries to peck out the back of the other bird's head. . . . They'll fight, but one bird eventually will try to run. In the wild it may escape but in captivity there's nowhere to run to. That's the problem.

"You find that with other species of birds," Derrickson said. "But it is difficult with rails because you can't see what's going on. They are very secretive."

At the slightest movement in the area of the cage, the birds skitter into hiding and it is next to impossible to find out what they are up to. "You have to use remote video cameras or a blind. . . . You are trying to watch them. They are trying to hide," Derrickson said.

"It's a fair problem. You have to watch these guys. We've even had pairs that have bred several times and then the male will suddenly turn around and kill the female or the female will turn around and kill the male.

"It isn't difficult to breed the rails. The most difficult thing is getting a compatible pair. That is still a problem with our whole program. You may want to breed this male with this female, but it's not so easy to do it. Getting them to form a good solid pair bond is difficult. They are just so aggressive."

As the chicks get older, the juveniles will turn on each other and on the younger chicks as well. "That doesn't happen in the wild," Beck said. "Obviously the captive breeding situation is in some way stressful. Maybe it is just that there isn't enough room for these young chicks to escape from these one-month-old birds."

The rail stud book is filled with violent deaths. There is, for example, the case of two male siblings that had caged together since hatching—first on Guam and then at Front Royal—without any problems and then after one year of living together one bird turned on and killed the other.

There have been some deaths due to bacterial infections, malnutrition, and other problems, but the vast majority of deaths, Derrickson said, have been the result of fighting.

At Front Royal, the rails are bred in a state-of-the-art bird building. Passive solar light filters down from skylights and into eighteen spacious pens filled with bunch grass, bamboo, and potted tropical plants. The floor is covered with a mixture of peat moss, volcanic rocks, and pine bark, and beneath that is a drainage system.

Each day, between 7:30 and 8 A.M., keepers clean the cages, check the birds, water the plants, and feed the rails a meal consisting of ground hardboiled egg (shell included), chopped kale, corn grubs, soaked monkey biscuit, millet, canary seed, canned dog food, bread, evaporated milk, honey, carrot oil, and wheat germ. The repast is garnished with a sprinkling of mealworms, crickets, and oyster shells.

Late in the afternoon the keepers are back to note how much the birds ate and once again check their condition. They enter all observations in "the green book," a fat, cloth-bound diary, which is the running daily history of the birds. Aside from these two intrusions—and the occasional keeper who slips into the big plywood box that serves as a blind and spends several hours observing the birds—the rails are pretty much left on their own.

Over time the curators have begun to cope with, if not fully understand, the rail's belligerence. Most of the violence has to do with sex. "You start putting two males together or two females together," Derrickson said, "and somebody is going to die." Similarly, putting together a male and female that are not disposed to mating will also lead to a death.

In an attempt to combat these tendencies, rail breeders began to place males and females they hoped to mate in the same cage, but with a screen between them. "We wait for the female to start laying eggs. That's a sure sign they are a compatible pair," Derrickson said. "Of course, the eggs won't be fertile. So we throw those eggs away and move the birds to a large pen. We put them in together and start hoping and praying the male doesn't kill the female."

As for the fighting among chicks, the problem is that these birds are remarkably precocious, reaching sexual maturity just six months after they are born. "So here you are holding youngsters together in a group, and they suddenly become sexually mature and usually one of the males is dominant and starts killing other males," Derrickson said. "So we have to split up broods.

"We have to sex them early," he said. "We try to do that by three months of age. We try to split them up into male groups and female groups. But then they can't be held in groups after six months of age any longer. They have to be held as individuals."

Similarly, as soon as the parents begin nesting again, the chicks have to be pulled. "You can imagine, you get a couple of pairs of rails breeding, how quickly, with that reproductive rate and that young age of sexual maturity, a population can grow," Derrickson said. One can also imagine all the sibling and spousal blood feuds that can be created.

Figuring out the sex of a three-month-old chick is not, however, as easy as rolling a puppy over on its back and checking its private parts. Derrickson and the veterinary staff at Front Royal had to develop a surgical procedure to do it.

"We sex them with a laparoscope, a fiber optic tube," he explained. "You make a small incision and you go in with a fiber optic tube and you actually look at the ovaries or testes to determine which sex they are. We had to test on a whole series of birds to determine when the best time was to do that. Around three months appears to be the best time to do it. You can be relatively sure you are not going to make a mistake."

All new chicks born in mainland zoos are sent to Front Royal within two months. "We sex them here and then they are maintained in quarantine, so I can turn around and send them to Guam or I can send them to another zoo for breeding. So there are a lot of shipments going in and out," Derrickson said.

This rail redoubt is housed in Front Royal's animal hospital. The laparoscopy is done in the small animal surgery and the rails are billeted in a set of rooms behind thick, sliding, metal doors.

All rails go through Front Royal at least once and some more often, as Derrickson puts together groups of birds that might be sent back to Guam for eventual reintroduction to the wild.

"It's nice to have birds breeding at a rapid rate and be able to produce numbers. But at the same time, it is really difficult to manage under those circumstances where you are having to move birds from one location to another and shift birds back to Guam," Derrickson said. "It is a never-ending job."

It is a constant effort keeping track of which zoos are hatching birds, arranging for the shipping of birds from zoo to zoo, and balancing activities between mainland zoos and the Guam rail facility on the other side of the world. And it is all driven by the fecundity of the rails. "There is no putting things off," Derrickson said. "It is frustrating."

What was been most frustrating, however, was trying to cope with the rail's aggressive nature and still mate the most appropriate pairs of birds to prevent the species from genetically collapsing in on itself.

The Guam rails, despite their fecundity, are facing a serious "genetic bottleneck." A mere twenty-one animals were pulled from the wild, and only fifteen of those survived to be adults. There were the three chicks from the same brood Beck, Morton, and Muna captured and three birds hatched from eggs out of the same nest. So the ability to mix and match birds was at the outset very limited. When the problem of violence and death among the rails was added to the equation, the bottleneck became incredibly tight.

The individuals within every species of plant and animal represent a range of characteristics—some are bigger, some shorter, some more mobile, some better looking. These traits are carried on the genes of these separate plants and animals.

All the genes from all the individuals form that species' gene pool. It is from this pool that the population emerges each generation. The generation refills the pool with its progeny and the species goes on into the future. The broader and deeper the pool, the more vitality the species has, the more ways it can respond to changes around it, and the more the pool can offer to the evolutionary flow of life on Earth.

The Guam rail's gene pool was now limited to just fifteen birds, some of them related, which restricted their likely mates. Not much to work with.

The first risk with such a small band of birds was simply not being able to get enough breeding pairs going. The next big risk was the prospect of inbreeding among those pairs. Individuals in the population may carry genes for injurious traits—defects or weaknesses—that are generally "recessive" or dormant, pushed to the deep, dark edges of the pool by the struggle to be fit and survive.

But when two animals with that gene are bred there is a greater likelihood it will surface and the prospect of this happening increases with increased inbreeding, as the work of Flesness, Ballou, and Ralls had shown.

But that wasn't the only threat the rails faced; there was also the menace of "genetic drift." In 1908, Godfrey Harold Hardy and Wilhelm Weinberg each independently postulated a theorem that became the basis for population genetics and was dubbed the Hardy-Weinberg principle.

According to Hardy-Weinberg, two gene alleles—the pair of genes contributed by the male and female—will have the same frequency in the gene pool, generation after generation. As a result, the number of times a particular trait will surface will depend on how many genes in the gene pool carry it. A common trait will surface often and be carried by a large number of individuals in the population. They will pass it on to their offspring and it will consequently remain common in the gene pool.

This will continue, according to Hardy-Weinberg, until some "agent" acts upon the gene pool to change the frequency. In the case of the rail, the agent was the brown tree snake, and there was now a danger that with such a small population, just by random variation, some genes might become more pronounced and others completely disappear.

Between the genetic bottleneck and the genetic drift, the rails had very little room to navigate on their odyssey of survival. The only thing they had going for them was that they certainly could copulate.

Derrickson, who was both the rail stud-book keeper and the species survival plan coordinator, began matching pairs of birds from different capture sites on Guam, under the theory that they were less likely to be siblings.

Breeders on just about every other endangered species project stopped there. But Derrickson went a step further, to the cutting edge of genetics research. He arranged for a young researcher, Susan Haig, to perform one of the most detailed genetic analyses ever done of a captive, endangered bird population using technologies such as electrophoresis and DNA fingerprinting.

Haig needed to plumb the rail's gene pool to see precisely what was left in it. First she attempted to determine the range of genes that the Guam rail had. She did this by electrophoretic analysis, a biochemical technique that can identify variations in proteins. Haig took rail blood samples for this analysis and sent them to Cornell University's Laboratory for Ecological and Evolutionary Genetics.

At the lab, the blood was placed in special gels and an electric current was run through it. The blood proteins separate at different rates, creating a pattern. When Haig reviewed these protein profiles, she found that the Guam rail had a broader range of genetic variation than some common rail species, such as the Virginia rail and the clapper rail. "That was the good news," she said.

While it was heartening that the Guam rail, as a species, had more genetic variation than the Virginia rail, the real question was how much variation there was among the fifteen Guam rails in the founder population. In other words, were these birds so closely related that they shared much the same genetic profile?

For an answer, Haig used DNA fingerprinting, which employs a radioisotope to mark fragments of a subject's DNA. The fragment is then illuminated by x-rays and the pattern captured on x-ray film. That pattern, a series of bands, is unique for every individual.

DNA fingerprinting has been most celebrated as a new foren-sic technique. It has been used in criminal trials, much the same way real fingerprints are. Increasingly, however, it is being seen as a tool in managing rare animal populations such as African elephants, California condors, and Galapagos tortoises.

The news from Haig's analysis—which was done on both blood and organ tissues—was also promising. Seven birds were not related at all. Each of the two sets of birds taken as chicks or eggs were, of course, related among themselves, but the sets weren't related to each other or the other seven birds. "We de-veloped a pedigree and a protocol for breeding based on the ge-netic identity of the founders and on an assessment of the best breeding strategies," she said. "The situation was about as good as could be expected."

Derrickson now keeps track of the birds on a chart, with the females ranged along the vertical and the males on the hori-zontal. For every two birds there is a calculation of how closely related they are.

Choosing one of the founders, he ran his finger along the axis. Most of the numbers were small. "This particular bird I can mate with a whole bunch of different males and not have an inbreeding coefficient above zero," he said. Nevertheless, as he began to match the birds with the more recently bred rails the coefficients began to rise.

Then he picked one of the more recent birds, Number 260. "I'm having trouble finding a mate that doesn't give an inbreed-ing coefficient. The longer you keep the population in captivity the more inbreeding you are going to have," he explained. "So what you have to do then is manage . . . by trying to form pairs based on the ones that give you the lowest inbreeding co-efficient."

In November 1989, Derrickson held another meeting at Front Royal to assess the state of affairs for the Guam birds. There were 150 Guam rails in captivity and the number of zoos in the program had grown to eleven. Guam rails were now to be found in San Diego, Tampa, and Kansas City. In all there were a total of eighteen zoos breeding rails, kingfishers, or both.

Instead of a half-dozen people, this meeting drew nearly forty and lasted three days. The goals set for the kingfisher were to increase the population to a rosily optimistic 200 by 1996 and add another ten zoos, bringing the number of participating institutions to twenty-five.

For the rail, the group set the goal of getting them back into the wild. But that was something that Bob Beck was already working on.

14

June 1986

Stuart Pimm was sitting in his office in Hesler Hall—the biology building on the University of Tennessee's Knoxville campus—when Bob Beck ambled in and introduced himself. After trapping all the birds he could and carting them to the mainland for captive breeding, Beck was back in Tennessee to take a stab at updating and finishing his two-year-old doctoral dissertation.

But it wasn't Smoky Mountain vireos Beck wanted to talk to Pimm about, it was Guam rails. The birds were breeding successfully enough that thoughts had turned to reintroducing them to the wild—somewhere. Guam itself was still out, but there were Cocos Island and Guam's neighbor Rota, as well as some small uninhabited islands in the Marianas chain to the north of Saipan. There were places to release the birds, and there were birds to release. The question was how to release them. That's what Beck wanted to discuss.

Pimm, an owlish and intense Englishman, was an expert on the issues of the dynamics of populations and communities. He had researched issues such as the structure and property of food webs—or more simply, who eats whom. He had pondered the question of stability and instability in ecosystems and the role of competition in shaping natural communities.

One of the subjects he occupied himself with was why and how populations vary and why small populations go extinct. If Pimm understood why small populations crashed, perhaps he might have some thoughts on how to ensure that small populations would grow.

The office that Beck entered was well ordered. The sun streamed through leaded glass windows and onto a well-waxed checkerboard floor of black and brown tiles. Scientific periodicals, such as the *American Naturalist* and the *Journal of Theoretical Biology*, sat in glass-door bookcases, boxed by year. Pimm's first cover article in *Nature*, "Complexity and Stability in Ecosystems," was framed. On the wall opposite his desk were a bulletin board and a yearly planner.

The bulletin board held photos—not casually placed, but arranged in an almost geometric mosaic—of various scientific gatherings and expeditions, including some fuzzy black-and-white shots from a trek through Afghanistan. That was Pimm's first field expedition, and one that left him so exhausted and sick that he missed an entire year at Oxford University, recuperating.

Now, with a glance at the yearly planner, Pimm could see that during the next twelve months, he was scheduled to visit Lausanne, Honolulu, Detroit, Mexico, and France. He had learned one great lesson in the deserts of Afghanistan—whenever possible he now flew first class.

A personal computer stood by the desk, its hard disk filled with articles, data bases, and correspondence. Next to the machine were framed photos of his parents in Derbyshire and his two daughters.

There were a couple of reasons Pimm seemed like a good person for Beck to look up. In addition to his expertise on the population dynamics of extinction, Pimm had done work on endangered tropical birds in Hawaii, and they shared mutual friends like Hawaiian biologist Sheila Conant. But perhaps just as important as his scientific acumen was the fact that Pimm had a reputation as a superb grantsman, and DAWR had no money to finance a reintroduction program.

Ironically, Pimm was already aware of Guam's problem because just a few weeks earlier he had reviewed a paper by Julie Savidge that had been submitted to the science journal *Ecology*. He had found the form of the article wanting but the material fascinating and suggested that the journal ask for major revisions. Now, here was someone in his office talking about the same island and the same birds.

"Bob and I hit it off because I had been working in the Pacific for a long time, and I was very interested in endangered species," Pimm said. "Bob had two needs. One of them was for expertise in how to go about doing a release, the population dynamics, how many, how often. That was an area where I had done some research. But he was also in need of someone to help him with the grant money, because of the peculiar way that things are funded federally."

Pimm was interested, but busy. To put a project together would take some planning, some traveling, some fund raising, and a graduate student to slog it out in the field. Pimm had not the time, the money, or the extra body. "So, we agreed to keep in touch," he said.

In August, Beck returned to Guam to take another stab at writing his thesis and getting the rails ready for freedom—somewhere.

As Beck left the University of Tennessee with the intent of writing his dissertation, Savidge arrived at the University of Illinois to defend hers. She had spent months writing the paper and now she faced the last hurdle. It was a slender document, just fifty-eight pages of text. Even adding the maps, charts, appendixes, and bibliography, it still ran only seventy-nine pages.

But that manuscript represented a good deal to Savidge. All the predawn races into the forest to raise mist nets. The nights walking the forest roads counting snakes. The lugging of batteries and cameras up cliffs, and the numerous attempts to build a snake trap, efforts marked by nicked and cut hands. It represented snake bites, now too numerous to count, and the hundreds of virological, bacterial, and blood samples taken from birds in the sentinel study. It was a compendium of three and a half years of Savidge's life.

It was also the distillation, the closing argument in the case against the brown tree snake as the perpetrator of one of the greatest avian extinction episodes of the twentieth century. And yet more than half the dissertation was spent putting to rest the specter of a Guam disease.

Between October 1982 and April 1985, a total of 380 exotic and native birds, representing forty-eight different species, had been surveyed for disease. More than 100 carcasses had been necropsied. No virus had been found, no unexpected bacteria revealed, only an occasional parasite, and no blood parasites at all were discovered in 233 samples. "The survey," the paper concluded, "revealed no evidence that disease was involved in the decline of the native forest birds."

The second portion of the paper recounted the sentinel study and its results, without going into Savidge's urge to flush white-eyes down a toilet. It pointed out that 106 birds, chickens, quail, canaries, and white-eyes were exposed to "flying vectors," i.e., mosquitoes and insects, for thirty-five to forty-nine days. One-third of these animals had their immune systems chemically suppressed. While birds did die, Savidge noted, the mortality rates of the protected and unprotected groups were similar. In addition, none of the deaths came as the result of a new or rare disease. "The study," Savidge wrote, "revealed no significant disease capable of causing the recent decline of the native forest birds on Guam."

The final section of the thesis opened with this blunt declaration: "The introduced brown tree snake *Boiga irregularis* is responsible for the range reductions and the extinctions of the forest avifauna on Guam."

Behind that statement Savidge arrayed her data on the expansion of the snake's range and the coinciding reduction in the range of the island's birds, the gut content analysis that showed the snake ate birds, the trapping study that demonstrated predation rates, the data on the decline of rat and shrew populations. She had laid out her three hypotheses and marshaled the data to prove them.

Now, she was back in Champaign-Urbana to defend her work. But Savidge already had plenty of practice fending off critics. Douglas Pratt, for one, had not gone quietly off into the night.

Back in 1985, after she had gotten her traps to work, Savidge was feeling pretty good. Then John Engbring visited the island

and showed her a draft manuscript of an article on the endangered birds of Micronesia that he and Pratt were collaborating on.

The article described the decline of birds on Guam and noted that the smallest went first. But instead of coming to the conclusion that they were easier prey, the authors concluded that might indicate some sort of breeding problem, since small birds tend to have shorter life spans and breed earlier and more often than larger birds. "Consistent with these observations is the possibility that that inimical factor operated not so much by killing the birds directly, but by preventing successful breeding," Pratt and Engbring wrote.

They went on to speculate that "such a pattern of die-off could have been caused by any agent (chemical pollutants, pesticides, radiation, disease, etc.) that rendered birds sterile, reduced viability of young, or interrupted breeding cycles, or any agent (such as a predator) that destroyed nests."

"I panicked," Savidge said, "because I was afraid that this was going to go into print." She telephoned the editor of the book in which the article was to appear. "I told him, 'Hey, you as an editor of this book have to confirm that what you've got in here is the truth,' and I pointed out some of the problems." But *Bird Conservation* was published later that year with the article only slightly modified.

In a review of the DAWR's proposed recovery plan, Pratt again fired a barrage of criticism at the disease work Savidge was doing. In a letter to Guam officials, Pratt said that "infectious disease is a much more likely cause of the rapid declines" than the brown tree snake and that disease research had been "inadequate."

Commenting on the design of the sentinel study, he wrote in a letter to DAWR chief Harry Kami that "the investigators obviously were interested in mosquito-borne disease." Pratt pointed out that since mosquitoes had been present on Guam for decades they were unlikely to be the carriers of the Guam disease. Then he asked why "contact and airborne pathogens hadn't been investigated.

"The circumstances of the bird decline clearly implicate disease; a negative finding in a search does not mean that no disease was present, only that none was found," he insisted. Pratt went on to ask whether any native birds had been directly exposed to known pathogens to determine their susceptibility in comparison with introduced species.

Beck, who was a knowledgeable, sympathetic, but somewhat detached observer, said years later: "Ruling out disease was a hard one. It's hard to prove a negative. Your critics can always say that you didn't test for all diseases, or it's a disease that you didn't think about, or it's a disease that is not in effect now. It is just very difficult to prove."

Savidge was off-island when Pratt's letter came, so Kami referred it to Sileo back in Madison. "Perhaps you may want to write to him directly," Kami suggested.

Sileo dutifully wrote a letter. He enclosed a copy of their summary report. He conceded that the report was incomplete and that the sentinel study had "deficiencies" as a result of having to be done quickly.

But he wrote: "I have been studying data and tissues from Guam since 1983 and I have yet to find evidence that disease was involved in the extirpation of the forest species. . . . I realize, as you point out, that this does not mean it is not there, only that it has not been found.

"But in twenty years of working with wild bird disease, I have never before faced an enigma like the Guam situation. We cannot even find a disease, let alone an agent that is causing disease," he wrote.

Sileo followed up the letter with a phone call. "I summarized our findings," he said. "I think he didn't believe me yet. I was just another incompetent federal researcher."

Pratt wasn't finished. In his field guide, *The Birds of Hawaii and the Tropical Pacific*, published in 1987, he wrote: "Many observers believe that circumstances implicate disease as the causative agent, but as yet no data have emerged to confirm or refute the idea. If diseases are involved, they are unlikely to be the same as those that decimated Hawaiian birds, because Guam birds have presumably been exposed to those pathogens and their vectors for a long time.

"Many researchers," he continued, "now think that the introduced brown tree snake (*Boiga irregularis*), a bird predator, is responsible for the Guam disaster. But this hypothesis does not, in our opinion, adequately explain the rapidity or thoroughness of the bird declines."

The only way to combat this, Savidge believed, was to get into print herself. "I wanted to get published as soon as possible, because I figured if it could survive a review and all of that, at least people would have the data available to make their own conclusions." Getting her fifty-eight pages past the dissertation committee was an important step.

They met in a conference room and sat around a table, four professors and one jet-lagged doctoral candidate. For more than two and a half hours, they grilled Savidge about the methods and results of her work and then they sent her out of the room.

The semester hadn't started yet and the corridor was empty. Savidge had been away so long that she probably wouldn't have recognized any of the graduate students in the department anyway. Then George Batzli, one of the members of the committee, came out of the room. "I remember it distinctly," she said. "He said, 'Congratulations,' and brought me back into the room and shook my hand.

"They said that I had passed and they would sign off on it. I remember I was so tired, really tired, and George looked at me and said, 'Julie, you should be smiling, you should be happy.' "

Savidge returned to Guam to complete a final draft of her thesis and revise the article she had submitted earlier that year to *Ecology*. Being published in a prestigious journal was as important to Savidge as passing doctoral muster. To be accepted for publication in a science journal, an article must survive a review by a jury of scientists.

Such peer approval would be another validation of her work, and would also present it to the scientific community at large. She had first submitted her piece to *Science*, the most distinguished journal in the country. But the article was too long, and one of the two reviewers raised some questions about the work on rats and shrews. "It was really a misunderstanding on his part of the data," she said. "But the way *Science* works, they won't let you respond, and you're out."

Next she had submitted the article to *Ecology*. By chance, one of the scientists asked to review the article was Pimm. As is his custom, after the children were in bed, the house quiet, and his own work concluded for the day, Pimm would pour a glass of wine and peruse review articles.

"I remember starting Julie's article and thinking, 'What is this?' She had written it in standard scientific form—an introduction, a section on methods and materials, results, and a discussion.

"The problem was, this wasn't a standard situation," Pimm said. "Yet within that paper was the most amazing story. You sat there and looked at what she had done and the arguments she had made, the clever experiments she had done. It was really a first-class piece of work, an uncut diamond."

The main reason the paper did not fit into the standard scientific format, which happens to work fine for laboratory bench experiments, is that Savidge's work covered much more than a hypothesis and an experimental test.

She had polled the people of Guam. She had used power outage data from island utility companies. She had reconstructed history from newspaper clips and DAWR field notes. "The idea of taking very different sources of information and bringing them together is the kind of thing that historians and people in the humanities are much more used to doing than scientists," Pimm said.

To be sure, Savidge had also done more traditional experiments, such as her predation study and sentinel study. But the scope of her work sprawled and did not fit neatly into the form developed by and for laboratory scientists.

Pimm wrote to the editors of *Ecology* saying that the paper needed to be rewritten, but that "whatever you do you've got to publish it." He also volunteered to write an accompanying essay, which he entitled "The Snake That Ate Guam."

Following Pimm's advice, Savidge revised the paper. The final version, "Extinction of an Island Forest Avifauna by an Introduced Snake," appeared in *Ecology* in 1987.

It read more like a story than a dry scientific treatise. It opened with a discussion of Guam and its disappearing birds. It then offered the list of possible causes and Savidge's three predictions. Then she recounted all her data.

The last portion of the nine-page, double-column article was a discussion of the results. Here Savidge argued that the massacre on Guam was aided by a set of singular conditions.

First, the snake had no natural predator and such elastic eating habits that the population could easily explode. Birds, on the other hand, appeared to have lower densities than on comparable islands, perhaps because of the simplified forest structure, which in turn was due to the repeated typhoons that raked Guam. It was also possible that the simple forest canopy made bird hunting easier for the snakes.

Taken together—a high snake population, a low bird population, and an easy hunting ground—the data made it apparent that the island's birds were in trouble.

"In summary," the article concluded, "the data clearly suggest that predation by the introduced snake *Boiga irregularis* is responsible for the decline of ten species of native forest birds, as well as several other types of birds on Guam. This is the first time a reptile has been implicated in the decimation of an avifauna, and the example shows how rapidly extinction can ensue under the appropriate ecological circumstances."

What Savidge had actually chronicled was the path of destruction by one of Edward O. Wilson's "horsemen of environmental apocalypse": introduced species. While the story of the first reptile responsible for extirpating birds was a new twist, Guam had been suffering, in one way or another, from this onslaught of alien species for centuries.

The weed overrunning Guam, which Savidge's husband, Tom, had toiled with, *Chromolaena odorata*, was another introduced menace to the island. So were the wild deer, feral pigs, and black drongos. Seibert's story ended, however, even more happily than his wife's. After several failures at establishing *Pareuchaetes*, he managed to populate the island with the little yellow moth, whose larvae ate the weed, and in a year's time the invading plant disappeared. "The native ferns came back, as if they'd been there just waiting," he said, "and pretty soon you couldn't find any *Chromolaena*." The pesky weed did not completely disappear, however. Small patches of it remained here and there, just waiting for another chance to take over. But it was held in check by the ever-hungry *Pareuchaetes* moth.

Guam's plight, however, isn't unique. The Hawaiian Islands have 4,600 species of non-native plants, three times the number of indigenous species.

The little partula snail is being bred in American zoos because competing South American snails reached the partula's South Pacific home and began to destroy it.

Nor is the problem limited to islands. "The threats posed by exotics are often understated, frequently dismissed as serious problems only for oceanic islands," contends Stanley A. Temple, a University of Wisconsin ecologist. "Island ecosystems have, of course, been extensively affected, but they are actually microcosms of what can happen to remnant continental ecosystems that are naturally islandlike, such as lakes, or that have become insular through the process of fragmentation."

In 1937, there were six exotic plant species in Kruger National Park in South Africa's northern Transvaal. A detailed survey in the 1950s recorded forty-three. By the 1980s, the number was more than 160.

In the United States, the bacteria *Borrelia burgdorferi* arrived from Europe, infected ticks, and created the Lyme disease problem. The zebra mussel came to the Great Lakes in bilge water from the Caspian Sea and spread into great beds that destroyed fishing areas and clogged water intake pipes. Japanese kudzu runs riot through Georgia.

This accelerated movement of plants and animals around the globe by humans has been called "one of the great historical convulsions in the world's fauna and flora."

In an article entitled "The Onslaught of Alien Species and Other Challenges in the Coming Decade," Michael Soulé described it as "a universal flood of exotics" and predicted that "the intercontinental traffic of exotic organisms, including their associated pathogens, will grow with the increasing mobility of people and commodities."

The ultimate effect, Soulé warned, is that the unique will be overwhelmed by the hardy. "The flood of exotics will tend to homogenize ecological communities, especially those subject to anthropogenic disturbance," he wrote.

European zebra mussels in North American lakes. Scottish gorse on tropical Hawaii. The Melanesian brown tree snake in Micronesia. Soon many ecosystems, from the temperate to the tropical, could be very much the same. Just as one can fly thousands of miles across the Pacific to land on a seemingly remote, tropical island, and still find fast-food drive-ins and malls, so too there soon may be no escape from the most persistent plants and animals—an ecological McDonald's effect.

And unlike the other horsemen—habitat destruction, hunting, and disease—there is no easy way of stopping exotics once they have taken hold.

"Most problems in resource use and pollution can be corrected," observed biologist Bruce Coblentz. "It takes little imagination to discern that ceasing an environment-offending activity will improve the situation. In other words, we can usually stop what we are doing and in time (maybe five years, maybe fifty, maybe 500) the problem will be corrected. . . .

"In contrast most exotics, once they become established are usually permanent in ecological time. In a few instances exotic organisms may be eradicated or controlled, usually at great expense and after severe environmental disturbance has already occurred, but most are uncontrollable and in comparison with human life span are in place in perpetuity."

In the fall of 1987, Savidge bumped into Pratt at the annual American Ornithological Union meeting. "Okay, I believe, I believe," he said, throwing up his hands.

"Eventually there were some real smoking gun examples," Pratt conceded. "There was, for example, a little patch of habitat that was surrounded by concrete runways where the snake had not gotten to, well as soon as the snake got there, the birds were wiped out. That's a smoking gun if there ever was one."

Ironically, one of the biggest stumbling blocks to the snake hypothesis—Paul Conry's work with turtle doves—eventually became yet another smoking gun. For after initially expanding into the areas vacated by native birds, the doves also crashed.

While the native birds were dwelling in the woods, Conry said, the Philippine turtle doves tended to inhabit the forest edges. As the snake wiped out the native forest doves it provided more habitat and food for the exotic doves. This led to what is known as an "ecological release"—an explosion in population. This was also probably facilitated by the fact that the Philippine dove laid two eggs at a time as opposed to just one for the Marianas dove.

But starting in 1983—just around the time that Savidge gave her first paper at the AOU meeting in New York City—the Philippine turtle dove population began to decline too, showing a 50 percent reduction over the next three years. The cause—nest predation.

Other introduced species also showed a decline in populations in the 1980s, although many of them also displayed more tenacity in hanging on in the face of the snake than the native birds.

"I've been beaten into submission by the facts," Pratt said. "But I think it was perhaps some of the skepticism that I and others expressed that made the research a little more rigorous.

"I never faulted Julie's work on the snake. What I say is she may very well be right. Probably is, given what we've seen subsequently. But she may not have the whole story. We may never have the whole story. The sad thing is we may never know. We are looking at something that happened twenty years ago. The trail is cold."

The story, Savidge conceded, is "not completely clean." But, she said, "science is never clean. Not for us out in the field."

Savidge stayed on Guam through the summer months, finishing up some work on the biology and ecology of the snake. She and Tom also had a little girl. But her work on the island was finished. Soon she would be on the windswept plains of Nebraska, teaching Cornhuskers and pondering the fate of the Platte River.

Pimm's work in Micronesia had, however, just begun, and the man with a yearly planner on his wall was about to meet island time.

15

M a y   1 9 8 7

Greg Witteman—computer hacker, biology graduate student, wrestler, and organist—was shuttling between conferences on conservation biology and nonlinear science at San Diego State University when he ran across Stuart Pimm. "I just sort of bumped into Stuart there in one of the conservation sessions, and we started talking, and he started telling me about the Guam rail project," Witteman said.

The twenty-five-year-old Witteman nursed the two seemingly separate interests of biology and computers. He had worked his way through college by building on-board computers for unmanned bathospheres at the Scripps Institute of Oceanography in nearby La Jolla. He was now at San Diego State doing a master's degree on food webs. "Theoretical stuff," he explained.

Pimm was in Southern California on a sabbatical. During the past few months he had thought more and more about Guam rails, and during his time in Southern California he had just about decided to help Beck with the release. Now before him was a strapping young man who could write computer software, had experience radio tracking birds, and, best of all, was looking for a Ph.D. dissertation topic. Pimm had found his able body.

"I decided there was nothing else for it. I knew I was going to become involved in the rail project, and if I was going to write a research proposal to get money for the release, I would need to go and visit Beck," Pimm said. One more scientist grabbed the red-eye to Agana.

Guam came as a shock. "I hated it," Pimm said. Part of Pimm's distaste was the result of the "avian desert" Guam had become. "It really is Rachel Carson's nightmare," he said. But part of it was engendered by the island's building boom that was plunking hotels down along the shore, plowing golf courses through the woods, and sprinkling fast food restaurants at every intersection.

Pimm and Beck spent a few days on Guam meeting with DAWR officials and discussing possible release sites. Several division and federal wildlife officials had strongly lobbied for doing the release on Cocos Island, the slip of an atoll just off Guam's coast. It certainly would be cheap and easy.

The two biologists went down to look at the island. When he got there, Pimm, who had diligently read and seconded Savidge's paper, was flabbergasted. "If there was ever a piece of evidence that it was the snake and not anything else, it was Cocos. The place was thick with birds," he said. Indeed, a variety birds, although none of Guam's key native species, still made their home on the tiny island.

Furthermore, birds and insects flew back and forth between Cocos and the main island all the time. It was highly unlikely that any disease affecting birds on Guam would not take its toll on Cocos. But brown tree snakes don't swim.

Pimm agreed with Beck that the atoll was too small and likely to be inundated during a storm. "It just wasn't the place to put a ground-dwelling bird," Beck maintained.

There were several uninhabited islands north of Guam in the Marianas chain—the weathered tops of a sunken volcanic mountain range. They would provide suitable habitat. The closest of these, Aguiguan, also known as Goat Island, was only forty-five miles from Guam, but there was no commercial boat or air service to the island, and as its nickname implied, the island had a feral goat population that might interfere with the introduction of rails.

The other uninhabited islands—such as Fallon de Medinilla and Anatahan—were more than 200 miles away from Guam by ship. Then there was Rota, Beck's prime choice.

Twenty minutes in an Air Micronesia twin engine propeller

plane and you were on Rota. "Air Mike" had not begun regular, twice-a-day service to Rota until 1988. But the island had a spiffy little airport complete with customs inspectors and a restaurant.

Rota sits just thirty-six miles north of Ritidian Point. But the islands are truly separate entities. So, when Pimm touched down, he found a place that was the antithesis of Guam. "The first thing you see when you drive down the road from the airport, if you are a biologist, is a lot of birds," he said. "But you come around a swooping bend and as you go down the hill, you can see the shoreline exposed, and it is absolutely breathtaking. God, what a lovely place."

What Pimm gazed upon were massive cliffs plunging to sandy white beaches fringed with swaying coconut palms and white surf. Rota sits like a big limestone birthday cake, surrounded by water so intensely blue, it appears to be lit from the ocean floor by huge neon lights. The base of the cake is ringed with beaches, coconut groves, and forests of banyan, breadfruit, papaya, and mango. Rising above these woodlands are sheer, green cliffs almost 1,600 feet tall. On the cake's flat top is a savanna of sprawling green fields and large stands of rain forest that have been growing largely untouched by humans since the island thrust itself out of the Pacific. "Rota is everybody's idea of a tropical paradise," Pimm said.

The two biologists drove the six miles from the airport to the village of Songsong, home to most of Rota's 1,300 residents. That six-mile stretch was the only paved road on the island, and it gave out just at the edge of the town, which was nothing more than a grid of dirt roads lined with one- and two-story stucco and cinder block buildings. A constant stream of auto traffic—nobody walks on Rota—churned up clouds of fine gray dust. There were no street signs, no stop signs, and no traffic lights.

The biggest building in town was the Church of San Francisco de Borja R.C., named after the island's patron saint, a former Spanish duke and sixteenth-century head of the Jesuits. It is a barn of a place with dozens of plaster statues, a linoleum floor, steel rafters lined with ceiling fans, and pews worn smooth and shiny by many years of prayer.

Pimm and Beck checked into the Blue Peninsula, a hotel above Penny's Meitetsu, the largest supermarket in town, where the aroma of soap and wax permeate everything, even the bread and vegetables.

The "Blue Pen" was no first-class hostelry. "It was really a dump," Beck said. "The place was so bad that after a day in the field, Stuart was in the shower and the water went off and he was all lathered up. I had to splash him down with some canteens."

"Sporadic water, sporadic electricity," Pimm said. "What can I say? It is what you'd expect a remote Pacific island would be like."

On such a tiny island—Rota was barely ten miles long and three miles wide—where so few travelers pass, this scientific odd couple, the tall, gangling Beck and the short, stout Pimm, were bound to be an event. "I remember the second day we went to the bar to have Spam and soba noodles, which was their specialty that day—every day—and people were really curious," Pimm said. "So we spent a lot of time talking to people about the rail project."

"We were looking for practical things," Beck said. "What kind of [logistical] support there would be for a graduate student, what apartments were available, what kind of food was around. We visited a number of sites and started formulating in our minds where we ought to release the birds."

It is remarkable that thirty-six miles of ocean could make such a difference, but islands are distinct not only in geography but time and history as well. As a consequence, Rota's story is entirely different from that of its bigger neighbor to the south.

The Spanish found the Chamorros living on Rota and other Marianas islands even more resistant to Catholicism and colonization than Matapan and his crew. They were forcibly removed in the late seventeenth century and resettled on Guam. They were not allowed to return for almost 200 years.

After Captain Glass seized Guam during the Spanish-American War, the Spaniards sold the rest the of their Micronesian possessions to Germany in 1899. The Germans' main interest in the islands was copra production and Guamanians were offered free homesteads.

A Japanese naval squadron captured the islands during World War I and ruled the Marianas under a League of Nations trusteeship during the years between the wars.

The Japanese developed extensive sugar cane plantations on Rota and built sugar mills and a railroad. Japanese became the lingua franca. Indeed, until the Japanese arrived the name of the place had actually been Luta, but the island's new masters changed it for ease of pronunciation. A 1936 census showed there were 4,729 Japanese, 787 Chamorros, and 48 Koreans on Rota.

In summer 1944, the American armed forces took the Marianas islands from the Japanese. There was a bloody three-week battle over Saipan, about seventy miles to the north. In an effort to buttress its forces, Japan sent three aircraft carriers. The United States had about sixteen carriers west of the Marianas, and they met the Japanese in the Battle of the Philippine Sea, also known as "the great Marianas turkey shoot." All three Japanese carriers were destroyed in a single day.

On July 6, the Japanese forces on Saipan launched a fanatical, last-ditch charge that did not end until there were 5,000 dead Japanese soldiers on the ground.

After securing Saipan, American forces invaded Guam and Tinian. They skipped Rota, which sits between the two, considering it unworthy of attack. Its virgin forests, pristine beaches, and majestic cliffs are inarguably the better off today for missing any intense fighting during the war. The various waves of occupation have, however, left their mark on the people if not on the terrain. The common street language today is a curious mixture of Spanish, English, Japanese, and Chamorro dialect.

During World War II more than 3,000 Americans died on the Marianas and another 110,000 were wounded. Never again would the United States display such an intense interest in the region. After the war, the islands were placed in a United Nations trusteeship overseen by the United States.

One American administration after another followed a policy of benign neglect. The CIA used Saipan to train Asian operatives, and the navy ran the islands until 1962. Then the Trust Territory of the Pacific Islands headquarters was moved to Saipan and the islands came under the auspices of the Department of the Interior.

But all was not well in paradise. A 1971 *Honolulu Star Bulletin* story reported that, frustrated by their status or lack of it, "the people of Saipan and the other Marianas have been moved to the brink of revolt. The Saipanese and their neighbors on Tinian, Rota, and the other Marianas Islands are unwilling subjects of the Trust Territory of the Pacific Islands, an American enterprise conducted largely for American interests."

To show their unhappiness, the islanders torched the "Congress" building, where the surrogate legislature met, and they threatened to secede.

The American approach to the discontent was twofold. First, negotiations were begun, which led to the creation of the Commonwealth of the Northern Marianas Islands, or the CNMI. Under a covenant with the United States, the islands became internally self-governing. The mainland, however, continued to control foreign and defense policy, and the military retained an explicit right to any land it needed for military installations.

On top of that, the U.S. government showered the inhabitants with federal perks—food stamps, Medicaid, subsidized housing, welfare, school lunch programs, and old-age assistance. There was also a hefty boost in government jobs. By the late 1980s, one-third of the CNMI was receiving food stamps, and Rota, an island with a population of 1,300, had a local government with forty different offices, agencies, and bureaus, including a Department of Naturalization and Immigration, an Office of Community and Cultural Affairs, and a program for the elderly. Government is the island's biggest business.

Presiding over all this was Prudencio T. Manglona, the mayor. A slow-talking block of a man, Manglona was not only the leader of Rota's government but the head of Rota's first family. His brother Benjamin was lieutenant governor of the commonwealth, his son Paul was the island's senator to the CNMI legislature, and his niece Laura served on the island's council. The Manglonas also owned Rota's local construction company.

Since Rota is populated primarily by members of only eight clans, it isn't surprising that family members turn up in elected or appointed positions. In fact, it seems that just about everybody on the island is related to everybody else.

The island didn't have telephones until 1986, but by 1990 there were about eighty private numbers. The Ataligs, Calvos, and Hocogs, all major families on the island, had six each. The Manglonas has eight.

In doing anything on Rota, even a rail release project, it was important to gain Prudencio Manglona's blessing. So Beck and Pimm made sure to stop by the little one-story house that served as town hall. "We went to see the mayor and explained the program to him. That was very important—not much happens on the island without the mayor," Beck said.

From a biological viewpoint Rota seemed an ideal place to release the rails. Sixty percent of the island's 22,000 acres was native forest, with another 36 percent open fields and scrub. Less than 2 percent of the island was developed. There were no feral pigs or goats, although there were feral cats and rats.

Still, there was a healthy avian population with sixteen different species. The cardinal honeyeater might be gone from Guam, but it still lived on Rota. The island also had its own species of collared kingfishers, bridled white-eyes, and rufous fantails. The Marianas fruit dove might have been wiped out by the brown tree snake on Guam, but it still flourished on Rota. But best of all, Rota did not have a rail.

"The question," Pimm explained, "was what the rail would do if it were placed on Rota. There is a great deal of information on the impact that endangered species have on native communities and for one reason or another birds tend not to have major impacts. Mammals usually do. Mammals on oceanic islands almost always do."

The other concern was whether the rail could establish itself on the island. Pimm and Michael Moulton had done a study of bird introductions on Hawaii and had come to the conclusion that a major factor in determining the success of an introduced species was the absence or presence of a similar species in the ecosystem.

The presence of a comparable bird meant that the introduced species would have to compete for food and nesting sites with an animal already better adapted to the habitat. Rota, however, had no rails.

"The niche was open. There weren't any rails on Rota. They wouldn't go head to head with a native species. There was nothing remotely like the rail on Rota. So the rail was likely to succeed," Pimm concluded. Since the island had its own species of kingfisher, Guam's Micronesian kingfisher would have to find someplace else to be freed.

Biologically, Rota was a go for the rails. But there were the politics and the bureaucracy to be wrestled with. At least four levels of government—from Rota to Washington, D.C.—would have to approve the project.

Manglona, while generally receptive, expressed some concern that if an endangered species established itself on private lands, this might create problems for landowners seeking to develop their tracts in the future.

Beck was confident that everyone's interests could be addressed and that he could get Rota, the Commonwealth of the Northern Marianas Islands, the territory of Guam, and the federal Fish and Wildlife Service to sign on to the project. Rota it would be.

One day, not long after returning to Guam, Beck heard that there was a crew with chainsaws and machetes cutting paths through the forest up at Northwest Field. He went and checked it out.

He found that five surveying paths, three to eight feet wide, had been cut through the woods. Upon further inquiry, Beck discovered that the navy intended to place three large antennae for a "Relocatable Over the Horizon Radar" system or ROTHR. "It is a backscatter radar," he explained. "Three transmitters from Tinian would bounce their signal off the ionosphere and bounce it back down to the Pacific. They'd be able to keep track of the area from Vladivostok to Camron Bay."

The system required three large receivers on Guam, and Beck calculated, looking at the navy's plans, that each antenna would destroy about 600 acres of forest. Those sites just happened to be the nesting and foraging ground for Guam's Marianas crow and fruit bats. It was also the last known habitat of the Micronesian kingfisher. The area was also listed as prime habitat in the DAWR's recovery plan for the rail.

Beck met with base officials and he pointed out that the birds and bats that used those woods were now protected by the federal Endangered Species Act. "I was told by one navy guy: 'This is a big national defense project. This is important for national defense.' But that's what they always say," Beck said.

True, there were no rails or kingfishers left in the wild, but if they were ever going to be reintroduced to Guam these government lands had to be preserved, because the island's building boom was wiping out suitable habitat up and down Guam.

"My opinion is that even if the snake hadn't come along by this time many of our species would be eligible for listing as endangered species because of habitat loss alone," Beck said. "So we have been trying to preserve the remaining native forests on Guam, much of which is in northern Guam." Since the animals that used the forest were protected by the Endangered Species Act, the DAWR asked the Fish and Wildlife Service to review the navy proposal.

Much to the DAWR's chagrin, the service ruled that the radar project would not jeopardize wildlife. The Guamanian officials were stymied. At this point the Marianas Audubon Society, the local chapter of the National Audubon Society, got involved.

Perhaps it was not surprising that Beck and a number of other DAWR staffers were active members of the organization. But they weren't sure what to do as environmental activists and the National Audubon Society could only offer moral support.

Beck was in San Francisco visiting family when he decided he would take a stab at interesting the Sierra Club Legal Defense Fund in his case. He called the San Francisco–based organization and spoke to Mike Sherwood, one of the staff attorneys.

"I just called him out of the blue," Beck said. "I couldn't tell you who I was representing when I went to talk to Sherwood. I just went to see him and we spent an afternoon talking. The next morning I got a call from him and he said: 'I think we can do something. We'd like to take the case.'"

Sherwood had practiced law in Hawaii and had an interest in both endangered species and island ecology. "Beck can be a persuasive guy, and he had stumbled onto a sympathetic ear when he stumbled onto me," the Sierra Club attorney said.

"It was a tragic story," Sherwood said. "Beyond that, I thought there was the chance for some precedent-setting law." One of the things that Sherwood found interesting about the case was that two of the key endangered animals—the kingfisher and the rail—could no longer be found in the forests but were still alive in captivity. "Usually, you are trying to protect the habitat an endangered animal is living in," he explained. "Here, we were trying to preserve the forest for the future for a time when the brown tree snake had been dealt with."

He believed that the Fish and Wildlife Service had failed to implement properly the Endangered Species Act and that the navy had failed to follow the provisions of the National Environmental Policy Act, which requires an extensive environmental impact statement before a government project is undertaken.

Sherwood began by writing detailed but amicable letters to both the navy and the Fish and Wildlife Service outlining the alleged breaches of federal law. The letters were followed up with a formal petition to the Fish and Wildlife Service to designate the forest as critical habitat.

That plea was buttressed by a second petition by the new governor of Guam, Joseph Ada, and supported by the island legislature. The ROTHR project had become a political cause célèbre on the island and Guamanians apparently had enough of the U.S. military deciding the fate of their land.

Slowly, the FWS set the wheels in motion to evaluate and designate critical habitat, and the navy went back and started an environmental impact statement. Bureaucracy being bureaucracy, several years passed. And then a funny thing happened: the cold war ended and the federal deficit ballooned.

"The ROTHR project was shelved. Congress simply would not fund it. It became a victim of history and economics," Sherwood said.

The Fish and Wildlife Service plodded on with the studies, federal notices, and public meetings to designate critical habitat. But another funny thing happened. Just as the final designation neared, politicians on Guam had second thoughts about protecting the cliffline and shore forests. Maybe there was no capital in a radar project, but a Japanese hotel and golf course? That was another story.

In the 1980s, Guam experienced yet another invasion. From Asan Beach, where 20,000 American marines hit the rocky shore, one could look across Agana Bay and see the new beach-head with its towering white monoliths. The Japanese were back.

This time the invading force was not soldiers but golf bag–toting tourists, and the point of attack was Tumon Bay with its picturesque white limestone cliffs, sandy beaches, and rainbow-colored reef. One after another they rose—the Sotetsu Tropicana Hotel, the Guam Dai-Ichi Hotel, the Fujita Guam Tumon Hotel, the Hotel Nikko Guam, the Guam Suehiro Hotel, and the Japan Plaza Hotel.

In 1976, about 206,000 Japanese tourists visited Guam. By 1987 it was more than 407,000, and in 1990 it would reach 780,000. "Guam became for Tokyo what the Carribean is for New York City," Beck said.

Only a three-hour plane ride from Japan's big cities, Guam, with a tropical climate and an American atmosphere, was coveted by Japanese developers. Land values soared and some Guamanians became rich. But the boom led to a level of building and economic activity the little island had never before seen and was ill equipped to handle.

Linda Sablon, a planner in the Guam Department of Commerce, keeps a tally of the island's progress on spread sheets. By 1990, Guam's forty hotels had 5,137 rooms and plans to add another 1,437. There were also proposals for forty-six new hotel and condominiums projects that would add still another 12,000 rooms.

"This has been spurred by the fact that we've had occupancy rates of 90 percent or more," she explained. "Guam tends to be the first place that a lot of Japanese try for vacation when they leave their country. It is a very big honeymoon spot.

"The Japanese are relatively unsophisticated tourists. They come on packages, go by bus, go to restaurants where they have contracts. They don't rent cars and don't explore much."

It is not only the sun and the surf that bring the Japanese to Guam, but the lure of arms and clubs. Japanese tourists display a great delight in shooting pistols and rifles—which are tightly regulated in their own country—and shooting ranges, some in a wild-west motif, dotted the Tumon area. Here for twenty dollars or so they blast away at targets with a Colt .45, a shotgun, or an automatic weapon.

But more than anything else it is the attraction of clubs—golf clubs—that excites the Japanese. By the early 1990s there were proposals to add twenty-three new golf links to Guam's three existing courses.

The Japanese love golf, but club memberships in Japan can run a quarter of a million dollars and greens fees can be a couple of hundred more. "It is cheaper to buy a condo on Guam and fly down three or four weekends a year than to buy a membership in Japan," Sablon said.

Along with Guam's prosperity came a new level of creature comfort. McDonald's, Pizza Hut, Taco Bell, and Safeway supermarkets opened and so did the Micronesian Mall, with its department stores and food court. Financed with Japanese yen, Guam was buying its American dream. The island's motto for the nineties could be "Guam—where America's malls begin."

The island became a land of opportunity for the entire region. From Palau, Truk, Pohnpei, and the Marianas, the young of Micronesia came seeking jobs and still Guam's unemployment rate remained less than 3 percent.

There has, however, been a downside to growth. Electricity, water, and sewer services have been severely taxed and the roads have become jammed with cars (in a single year there were 29,000 new auto registrations). Similarly, schools are overcrowded and apartments have become scarce.

By 1991, there were scheduled power blackouts moving from neighborhood to neighborhood to conserve electricity, and daily notices of which beaches were closed due to sewage pollution. Periodically the demands on the public water supply system became so great that the water simply ran out. "All the hotels are required to have their own desalinization plants," Sablon said, "so when there is a water problem we all go over to the hotels."

Infrastructure problems and the loss of nature habitat have not been the only woes visited on the island by the economic boom. Some social analysts warn that the waves of American federal funds, Japanese investment, and immigrants from other islands are slowly drowning traditional Chamorro culture. "We our losing traditions in a wash of federal aid," wrote Laura M. Torres Souder, a University of Guam sociologist.

Ultimately, the boom forced a measure of self- reflection on the part of Guamanians. A land use plan—the *I Tano'-Ta*—was drafted after public meetings across the island, and a debate rose about the island's political future. Many now believe that a more independent course—Guam for the Chamorros—is the best way to preserve both the island and its culture. "There is more interest in creating a commonwealth now, moving away from the United States a little bit to get more control over the island," Sablon said. "A commission for self-determination has been established. It is overwhelming, what is happening."

Money doesn't only buy dreams, it corrupts. Japanese developers with multimillion-dollar investments at stake have not been above currying favor with the island's ruling clique—both legally and illegally. In 1990, after a wide-ranging federal investigation, thirty officials were convicted of charges of taking kickbacks and illegal campaign contributions. Among the guilty was former Governor Ricardo Bordallo, the man who had personally telephoned Julie Savidge in an effort to dissuade her from worrying about brown tree snakes.

Ironically, Guam's resort land rush posed a far more serious threat to wildlife than the proposed ROTHR project, and probably the biggest threat since the coming of the brown tree snake. Just one of the proposed forty-six new projects—the MDI Guam Development Corporation's plan for 3,000 homes, a 200-room hotel, and a forty-five-hole golf course—would cover 1,350 acres, almost as much land as the navy's radar system.

And unlike the military projects, which the local government found easy to object to, it has been more difficult to say no to the tourist industry, which pumps almost $1 billion a year into the local economy.

As in the Amazon, the wetlands of the American mainland, and the rain forest of the Malay peninsula, the threat of habitat loss—the prime cause of the loss of species worldwide—had finally come to Guam.

Upon his return to the University of Tennessee from Rota, Pimm already had his war plan drawn up. "At that stage of the game, I started hustling money," he said. On his list of possible funding sources were Wildlife Conservation International, which is the conservation wing of the New York Zoological Society; Wildlife Preservation Trust International in Philadelphia; and the International Council for Bird Preservation."

Pimm sent proposals off to all three. He wasn't being greedy. He knew that a typical grant from one of these organizations was no more than $20,000, and by his calculation the rail project would need twice that for the first year. "So basically, we held our breaths for six months waiting to see if we would have any money at all to get the project going," he said.

World Conservation International eventually awarded the project $15,400, Wildlife Preservation Trust International anted up $10,000, the International Council for Bird Preservation contributed $5,000, and the World Wildlife Fund added $2,000. Pimm was delighted. "In actual fact, from any one of those organizations, we were lucky to get anything at all," he said. "It was a really high-risk kind of venture."

In the project proposal, Pimm carefully outlined the planned releases, the intent to radio track the birds. Each animal would carry a little backpack transmitter, and Greg Witteman would march across Rota following the fate of the birds. "Ideally, one day we will come across a male and female and their brood and there will be no radios on any of them," Pimm said.

He also set a target date for the release. Contingent upon funding, Pimm proposed releasing the birds in summer 1988, during the rainy season, when it was believed the animals could more easily adapt to the island.

In early 1988, with the grant money in hand, Witteman joined Pimm in Knoxville. While Beck handled the political and bureaucratic details in Micronesia, the two sat down to ponder how best to release rails into the wild.

Chapter

16

December 1987

There was a milling, seething, babbling mass of 200 children gathered in the grade-school cafeteria in Kolonia, Pohnpei. None of them had ever seen a snake, for there were no snakes on the island of Pohnpei. But now there were rumors that snakes had been seen on this Micronesian island, 750 miles east of Guam. Between the news and the wave of energy created anytime a couple of hundred children are placed in a room, on Pohnpei or anywhere else in the world, the hum in the cafeteria was straining to become a roar.

Before this motley assembly stood Gordon Rodda. This was definitely going to be a tough crowd, much tougher than college undergraduates. Still, it was a job. Just eight weeks before, Rodda had been a nominally unemployed former Smithsonian Institution postdoctoral fellow. Now he was about to explain what a snake was, what it looked like, and the risks to Pohnpei should the brown tree snake ever get there. Surely there were easier ways to earn a living.

Rodda had been studying sex and aggression among green iguanas on the broad savannas near Calabazo, Venezeula. But as so often happens, there was more research to do than money to do it. His Smithsonian fellowship ran out and he was forced back to the States, where he landed a part-time teaching job at the University of Tennessee. The teaching post paid the bills, while Rodda tried to complete his analysis and manuscript on the love life of the green iguana.

He had observed that the iguanas' mating ritual depended upon a form of sexual warfare in which males tried to force themselves upon females and females even fought among themselves for the right to stay in the best territories with the dominant males.

"There appears to be some sexual selection in both sexes simultaneously. The idea is that each sex is fighting among itself for access to the other sex," Rodda explained. He saw something "intuitively plausible" watching the sexual battles among the lizards. Certainly, there are scores of animal mating rituals where males vie for dominance and the right to territories or harems. But should it be any different for the females?

So there was Rodda writing up his iguana work and teaching part-time when he heard at the annual meeting of the American Society of Ichthyologists and Herpetologists that Tom Fritts was looking for somebody to work on snakes in the Pacific.

Rodda called. Fritts told him to send a résumé. He did. It showed that Rodda had graduated summa cum laude from the University of Colorado, was a member of Phi Beta Kappa, had a Ph.D. in animal behavior from Cornell, having done his dissertation on alligator navigation, and was a member in good standing of the "herp society."

Fritts hired him and within two weeks Rodda found himself on a plane flying across the Pacific. The reason for the urgency was reported snake sightings on a couple of islands, including Pohnpei.

Another island, still another reality. Pohnpei's story once again is totally different from that of Guam or Rota. Located 2,250 miles west of Hawaii and 750 miles east of Guam, Pohnpei is an island of soaring, limestone-capped volcanic peaks. Mount Totolom is nearly 2,600 feet high, twice as tall as Guam's tallest mountain. It is also one of the wettest places on Earth. It rains about 300 days a year. Kolonia, the major town, gets 172 inches annually. But the mountains get more than double that amount. The water comes racing down the slopes, leaping into silver waterfalls, diving into deep, clear, forest pools. The rain and rich volcanic soils create one of the most luxuriant coats of tropical foliage in the Pacific and give the island its nickname, "the garden of Micronesia." The 129-square-mile island, about twice the size of New York City's Staten Island, is surrounded by mangrove swamps and barrier reefs.

Politically, Pohnpei's story is also substantially different—right down to the the present. The Spanish perfunctorily claimed it in the sixteenth century, but never settled the land or even sent missionaries.

The first big Western impact came from American whaling ships. In 1854, the *Delta* stopped at Pohnpei and left smallpox in its wake. The disease wiped out half the population. A few years later, another whaler, the *Pearl*, brought measles to the island and claimed many more lives. The population of Pohnpei, indeed that of many of the Micronesian islands touched by these diseases, has never fully recovered. Today the island is home to about 24,000 people.

Pohnpei passed to the Germans in the same land purchase that brought them Rota in 1899. But unfortunately it experienced a different fate in Teutonic hands. The German administration instituted forced labor programs for public works projects, and in 1910 four German officials were killed in a dispute with members of the Kawath clan. In retribution, seventeen islanders were executed and 426 exiled to the phosphate mines on Ngeaur Island, more than 1,200 miles away.

The Japanese took over Pohnpei after World War I and adopted much the same development program they did on Rota, building up a population that outnumbered native Pohnpeians three to one. The island didn't figure much in the fighting during World War II. The Americans bombed four airbases but saw no need to invade. After the war, the island passed into the Americans' trust territory.

Today, Pohnpei is not a territory like Guam or part of a commonwealth like Rota, but a member of the Federated States of Micronesia. The federation includes four island groups—Pohnpei, Kosrae, Truk, and Yap—with forty inhabited islands.

These islands are spread across more than 1,500 miles of ocean but altogether have only 270 square miles of land and 73,000 people. Kolonia serves as the capital.

The federation was created under a "Compact of Free Association" with the United States in 1982. The deal gives the U.S.

military any land it needs, the right to store nuclear weapons on federation soil, and free passage for nuclear submarines and all American ships. In return, the federation is to receive $1 billion in aid over the fifteen-year life of the compact. That comes to more than $137,000 for every person on a federation island.

Pohnpei and the other federation islands have remained off the beaten track. American tourists never make it past Hawaii, and most Japanese tourists stop at Guam. When Rodda and Fritts touched down in Kolonia, their bags were deposited on the tarmac. They picked them up and carried them through the one-room corrugated iron shed that served as the customs house and airport terminal, and they were officially in Pohnpei.

Rodda was trying to show slides of snakes on a tiny screen at the front of the cafeteria. But there were no shades, and so there was no way to make the room dark enough for the slides to be seen. "Tom was amused," Rodda said, and Fritts agreed. "It was a pretty funny scene."

While "public education" was part of their mission—Fritts also held meetings with island officials and customs inspectors—the main purpose of the visit was to run down the snake rumors. Like Joe Friday and Bill Gannon, Fritts and Rodda went around the island looking for "the facts."

There had been four reported sightings of snakes. "Two of them we couldn't get anything on at all," Rodda said. "You've got to remember that the animal simply wasn't in these people's collective experience, so if they saw something they didn't know what they were looking at."

A third snake had been found in a schoolyard. The two herpetologists were able to identify it as a bronze back snake. The serpent is common on Palau, another of the Micronesian islands, and it was found near the community college, which has a number of Palauan students. Fritts suspected that the snake might have been part of a practical joke.

Still, even if it had been an invader, it posed little threat to the island's sixteen species of birds—which included four birds found nowhere else—the Pohnpei lory, the Pohnpei flycatcher, the Pohnpei greater white-eye, and the Pohnpei mountain starling. The bronze back simply didn't grow very large and existed largely on a diet of lizards.

The fourth snake had slithered out of a shipment of timber

that had come to Pohnpei from the Philippines by way of Guam. The lumberyard owner had killed the serpent. It turned out to be a wolf snake, so named for its enlarged canine-looking teeth. The snake, which like the brown tree snake is a rear-fanged colubrid, had apparently climbed aboard in the Philippines, stayed in the pile during the layover in Guam, and ridden all the way to Pohnpei.

The wolf snake, which comes from Southeast Asia, had a long history of hitching rides in cargo. That's how it got to the Philippines and established itself there sometime in the last century. As far as Fritts and Rodda could determine, the wolf snake sighting on Pohnpei was an isolated incident and the island was still safe.

But the wolf snake is still hitching rides and still colonizing islands. For example, Fritts had received reports that it had landed and established itself on the Australian territory of Christmas Island sometime in the late 1980s.

A fifty-two-square-mile blip located in the Indian Ocean, about 224 miles south of Java, Christmas Island is the top of an ancient ocean mountain. It has thick forests, steep cliffs, and coral and sand beaches. The island is home to seabirds, reptiles, and insects. There are also rich phosphate deposits.

There are no indigenous people. Most of the island's 3,000 residents live in the settlement of Flying Fish Cove and work for the Phosphate Mining Company of Christmas Island.

Around 1988, the first wolf snake was spotted, probably having arrived in cargo or with workers from Indonesia. Now, it is undeniably established on the island. "It doesn't grow as big as the brown tree snake, so it isn't an immediate threat to any larger animals," such as birds or small mammals, Rodda said. "But it is definitely there."

On his very first visit to Guam, Fritts had postulated the theory that brown tree snakes were so numerous that it was likely that they could get into military or commercial cargo and be transported to other islands, where they would establish colonies and wreak ecological havoc as they had on Guam.

He had shown that the snake could be found around cargo areas, and now he was slowly proving that it was getting off the island. He found his very first piece of evidence on his trip back to the mainland in June 1986. Fritts was carrying several brown tree snakes, and when he arrived at the Honolulu airport he immediately went to the customs area to declare the animals to a state Department of Agriculture inspector. "I told him I was transporting some snakes, some *Boiga irregularis*, and he wrote it on the form without even asking me how to spell it," Fritts said. "I asked how he knew about *Boiga* and he said, 'Oh, we had one up at Hickam Air Force Base last week.' "

Fritts parked the snakes with the agriculture inspector and went to Hickam. The specimen he saw there was definitely a brown tree snake. It had been discovered, by the grounds-keeping crew, slithering along the bank of a stream behind a maintenance shed near a transit aircraft parking pad.

Returning to the airport, Fritts spent some time in the Department of Agriculture's quarantine office, which was filled with specimen jars of animals and plants confiscated at the airport. There among the glass containers was a snake that had been found in the customs area of the airport back in 1981. It had been identified as *Boiga kraepelini*, but Fritts knew immediatley that it was another brown tree snake. The animal had undeniably already been able to make its way the 3,000 miles from Guam to Hawaii.

During the next year, Fritts collected incidents of brown tree snakes making their way to Kwajalein, which is just east of Pohnpei, and to Diego Garcia in the Indian Ocean.

There were also reports of snakes seen on Saipan and Tinian. In September 1984, the South Korean vessel the *Chung Ung* was being loaded with crushed car bodies at Apra Harbor, when crew members found and killed five snakes in the cargo.

Fritts's theory wasn't a theory any longer.

Still, getting money to work on the brown tree snake problem wasn't easy. Fritts had to figure out some way to simultaneously study the snake and provide data that funding sources considered valuable. The solution lay in the constant power outages the animal was causing on Guam.

Snake-induced power failures had become a way of life on the island, and 1987 had been a particularly bad year with a record ninety-five outages.

The snakes were climbing guy wires and dropping from trees and cliffs and every time one hit a live wire the animal became a conductor. When it touched a second wire or a ground, such as a pole, *bam*—there was one fried snake and one power outage. It appeared that the snakes were lured by sparrows that made their nests on poles. But part of it was simply the arboreal nature of the animals.

The blackouts were exacting an economic toll. The Guam Power Authority estimated that a single power outage in October 1986 had cost $15,000 in repairs and $60,000 in lost revenues. A fluctuation in power caused by the snake in April destroyed another $20,000 in equipment.

And those costs were being multiplied at businesses all over the island. Concerns that depended on computers, such as banks and the duty-free stores, were particularly vulnerable. But so were private homes, where power drops and surges burned out toaster ovens and air conditioners.

Earlier in the year, a snake-induced outage damaged one of the main generating facilities, lasted twelve hours, and cost the power authority more than $250,000. Following the blackout, there were 132 monetary claims filed with the authority for domestic appliance and commercial equipment damage.

All over the island, businesses were forced to cope with the outages. The duty-free store invested in a $30,000 backup battery. The Hilton Hotel, at first, placed candles and matches in every room and eventually increased the power of its own generators. In fact, lots of stores and offices bought their own generators.

In an effort to mitigate the impact of the snake, the power authority in late 1986 installed new circuitry designed to localize outages and prevent the entire island from being pitched into darkness. The Naval Public Works Command, which operates generators for military bases, actually shut down one of its 135-kilovolt lines every night because the line had suffered more than a hundred outages. A single pole had been hit by snakes seventeen times. Nevertheless, between 1978 and 1990, there had been nearly 1,000 electrical outages caused by snakes.

The Guam Power Authority tried to erect some defenses. It placed stainless steel collars around the fat concrete poles and screen disks, about three feet in diameter, on the guy wires. The guy wires were also coated with adhesive paints. But the snakes kept coming.

Considering the havoc they caused, Fritts suggested that one step in developing ways of protecting the electric lines was to identify how snakes are able to reach the top of power poles.

"I really wasn't interested in power poles, but it was a way of studying the snake," he said. It was also the kind of real-life problem that government agencies could relate to. The Department of the Interior and the Fish and Wildlife Service both kicked in some money, and the Guam Power Authority contributed $5,000.

Fritts had rounded up nearly $70,000 in funding in addition to his own salary and general administrative costs, which the FWS had agreed to pick up. One of the expenditures he made was for an additional researcher—Gordon Rodda.

The first analysis done on the electric problem was a collaboration between Fritts and Savidge based on Naval Public Works Command records from 1978 to 1985. The data showed that outages were more likely between 8 P.M. and 4 A.M., with ten to midnight the peak period. Outages were also most common from May through August—the beginning of the rainy season. These data offered not only a profile of power blackouts but also some insight into the activity of the snake.

Fritts began studying the snake's methods of locomotion. He found that the brown tree snake was a remarkably artful climber, adept at distributing its weight, controlling its center of gravity, and focusing all its force on the smallest of pivot points. Just a small bump or a rough spot was enough of a hold for the snake to glide up what appeared to be a smooth wall, making it appear as if the laws of gravity did not apply to the animal. He also learned that the snake was strong enough to hold as much as three-quarters of its body upright, so it could search for that pivot point, or stretch across a chasm from branch to branch. This meant that a five-foot snake would easily be able to bridge a gap of more than three and a half feet.

The Power Authority's defenses were failing, Fritts discovered, because they were ill conceived. The snake was not climbing the fat concrete power poles. Indeed, it could not climb the smooth poles, so the stainless steel skirts were useless.

As for the disks on the guy wires, they were ineffectual. Fritts found that the snakes would literally fling themselves over the contraptions. His analysis showed that the snakes were mainly using the guy wires and, to a lesser extent, nearby trees to reach the electrical wires.

Fritts recommended removing all the guy wires possible, lowering the remaining wires so the snakes couldn't reach the power lines, and cutting back vegetation surrounding the poles. The Guam Power Authority between its new circuit breakers and a program to trim back vegetation did reduce the severity of the outages. But Fritts added, "The last time I looked the guy wires were still up." And Guam still suffers occasional brown tree snake power outages.

After their visit to Pohnpei, Rodda and Fritts went to Guam. "Tom and I spent almost a month on Guam collecting snakes, talking to people, recording power outages, and getting a broad view of the whole situation, as well as getting to see a little bit of the less-developed parts of Micronesia," Rodda said.

By this time Fritts's research was beginning to win converts and money. In late 1987, the air force and navy came up with a total of $85,000 in funding for a three-year study aimed at preventing the spread of the snake in military traffic and cargo. Then when the furor broke over the ROTHR project, the navy asked Fritts to do an evaluation of the forests on Northwest Field.

"The folks over at the DAWR were furious. They felt like we'd betrayed them, that we'd gone over to the enemy. But all we were doing was telling the military what was there," Fritts said.

Fritts and Rodda surveyed the proposed sites and tripped across yet another baffling display of behavior by the brown tree snake. At one study site, the snakes had come down from the trees and were foraging and hunting on the ground. Apparently, they had wiped out all the arboreal geckos and were now busily hunting skinks on the forest floor. "It was totally unexpected," Fritts said.

The electric power pole work had shown that *Boiga irregularis* was an unsurpassed climber and master of the forest canopy, but the snakes at the Naval Communications Station showed themselves to be just as adept moving and hunting on the ground. The brown tree snake, it appeared, could go anywhere and do anything on Guam.

The exhibition also prompted Rodda to ponder the fate of the geckos. Were they, like the birds and the shrews, going to be wiped out?

Rodda also went back and reevaluated Savidge's and Fritts's trap data. "They had both been very conservative in estimating densities, basing their calculations only on the snakes they caught," Rodda said. Savidge had used thirty-two traps and calculated 6.5 snakes per acre. Fritts had used eighty traps and estimated that there were up to twenty-five per acre. But what about the snakes that had escaped or never been caught?

Rodda ran his own trap grid. But he did not keep his catch. Instead, each snake snared received a number on its head, in white indelible ink, and was released. After a few weeks, a ratio of new captures to "recaptures" emerged and that in turn gave Rodda an indication of the size of the population out in the woods.

When he applied these ratios to the previous trapping data, he concluded that in 1985 there had not been 6.5 or 25 snakes per acre, but nearly 30. His most recent trapping appeared to indicate that the population might have fallen to 14 snakes an acre.

Still, at these densities, save for a few wetlands, coastal stretches, and some dens where certain species hibernated, Rodda believed he was looking at the greatest density of snakes anywhere in the world.

It was inevitable that islanders and tree snakes would begin crossing each other's paths. As the years passed, the stories grew. There was the woman who was driving down a road when a small serpent shot out of the air conditioning vent and almost caused her to crash into a tree.

There was the man who went to change the oil pan on his car and was bitten by a snake that had been sleeping on the axle. A kennel owner found the carcass of a newborn puppy tangled in a wire fence. Apparently, a snake had killed the dog but then had failed to get its prey through the wire mesh. Again, it was curious that the snake seemed to know there was a litter at the kennel.

A sales clerk found a snake hanging from a clothes rack in a fashionable boutique in the Tumon Bay hotel district, and there were reports of snakes entering homes through plumbing pipes and ending up in tubs and sinks.

One woman was bitten on her behind as she was sitting on a toilet by a snake that came up through the pipe. The naval customs officials also had story about a rather large woman at a party in an officer's home going into the bathroom, seeing a tree snake, fainting, and wedging herself in so tightly that it was impossible to open the door. The fire department had to be summoned.

The brown tree snake had wormed its way into the economic and social life of the island. This was not, however, the script for a horror movie. Guamanians did not need flamethrowers to blast a path through snakes in an attempt to get to their automobiles each morning nor did they take turns standing guard through the night. There was no panic. It was like the typhoons that occasionally hit the place, just part of island life. Many of the people, like Doug Pratt, had never even seen a brown tree snake.

It was during this period that Bob Anderson passed on to Fritts a note he had received from Guam Memorial Hospital. At 6:50 A.M. the day before, a two-month-old boy had been admitted for a snake bite. He had been found in his crib with a five-foot snake wrapped around his neck and left arm. The serpent had been chewing on the baby's arm.

The boy was having trouble breathing and was treated with

norepinephrine. The infant responded and after two days in the hospital was released. Fritts found the incident curious. He tried unsuccessfully to find out if the respiratory problems were a result of shock, trauma, or the snake bite. The venom of the brown tree snake wasn't thought to be toxic, but then again in 1975 the distinguished herpetologist Robert Mertens had died from the bite of an African bird snake, another rear-fanged colubrid, and that wasn't supposed to have been dangerous either.

Chapter

17

Spring 1988

Once the necessary funds were secured, Stuart Pimm turned his attention to an entirely different numbers problem—how many rails should be set free on Rota and how often should the releases be done? Although this might have seemed like a rather esoteric, technical question, for Pimm it was a fundamental problem in population biology and possibly one key to saving endangered species.

While Pimm was no stranger to field work, having done his time in deserts and swamps, his reputation rested largely on his ability as an imaginative theorist, and in the Guam rail project he saw an opportunity to turn theory into practice. In addition, it offered a way to break through the traditional approach and formal conventions of most ecological studies.

Although the layman might assume that ecologists deal with the big environmental issues of our time—global warming, extinction of wildlife, the loss of natural habitat—in truth they deal with issues much, much smaller.

Ecologists are just like their scientific brethren. They are trying to understand and explain the world by reducing it to simplified models in which a limited number of variables can be measured and manipulated.

The result is that most academic ecological studies focus on a small group of species, in small areas, for relatively short periods of time.

When an ecological study of this sort is finished, one may know a good deal about mussels in the splash zone of rocky intertidal pools or insect propagation on the arctic tundra, but not much more.

Compounding this problem, at least from Pimm's viewpoint, is the fact that ecologists tend to do these studies not where environmental problems occur but rather in remote, unspoiled corners of the world.

"Most ecologists take themselves off to very beautiful places and they study nature in her pristine state and they study the wonderful and bizarre things that nature does. What they avoid, like the plague, are places where things are going wrong," Pimm said.

In the 1980s, a number of scientists began to try to construct a bridge between the ecologist laboring over a few square meters of rain forest and global environmental crisis by creating a new discipline—conservation biology.

The idea was to marshal scientific insight on issues of ecosystem operation, population dynamics, genetics, and related fields for use in applied projects to restore and preserve species and habitat.

This bridge between classical ecology and conservation is a tenuous one. The scale and time-frame of conservation problems are far greater than traditional studies. For example, even before starting his calculations, Pimm knew that there was a possibility that the Guam rail reintroduction program could take decades to reach success.

There is also a certain do-or-die imperative in conservation projects that academicans rarely have to face. "If you do conservation biology badly species will become extinct. That forces upon you a special kind of discipline," Pimm said.

Now, Pimm was about to embark on an exercise combining his theoretical expertise in population dynamics and his desire to set something right in the world.

Like Savidge, Sileo, or Fritts, he was a biologist seeking answers. But his research would need no mist nets, no traps, no pipettes of birds' blood. Instead, numbers, a computer, and some mathematical constructs were the required tools. He would conduct this search through the dynamics of avian populations in the comfortable confines of his office.

A "population," for a biologist, is a geographically discrete assemblage of individuals from a single species. One of the reasons biologists like to work on islands is that populations are so conveniently drawn. No overlap. No messy migrations. No interference. A neat, identifiable group.

One of the fundamental assumptions of population biology is that given the opportunity and proper environment, any population will just continue to grow and grow. For example, one study noted that in 1937 half a dozen ring-necked pheasants were placed on Protection Island, off the coast of Washington State, and five years later there were 1,325 birds on the island.

As a result, most of the discipline has been focused on understanding the forces that keep populations in check: the relation between birth rates and death rates; the interplay between increasing density and declining growth rates, for as a population becomes more plentiful, basic resources, such as food, are spread more thinly, and that in turn can affect fertility, thereby slowing growth.

Similarly, as prime territories are taken up, those without them—like Rodda's leisure-suited male green iguanas—lose out in the effort to reproduce. In some cases, those without such territories are more exposed to predators, which also help to keep populations in check.

Much study has been devoted to the relation of predators and prey and the cycles of ascent and decline the two populations go through, for as a predator wipes out its prey, it shifts the advantage from the hunter to the hunted. The survivors suddenly have more resources at their disposal and reduced numbers may make it harder for the predator to find them. If the prey population falls, invariably the population of predators soon follows, and if the predator reduces the prey base below some vital minimum, it is the predator itself that is in trouble.

All this had a certain relevance to the brown tree snake. Like many invading species, when it got to Guam, it found a virgin territory and experienced an ecological release as its population exploded.

Usually, the release ends at some point as a new equilibrium is established. The problem on Guam was that the snake population kept going until there were perhaps two million snakes.

The species' appetite turned out to be so eclectic that even when it had wiped out birds, rats, and shrews, it dined on skinks, geckos, and spareribs. The real prey base, Rodda's work was beginning to show, was other reptiles, which reproduced at an even faster pace than the snake.

But none of this directly helped Pimm, who was dealing with a very small population that might make it—like the ring-necked pheasant—or might not. His concern was not the dynamics of growth but the threat of extinction and how best to escape that threat. This was the question he wanted to answer.

"If we have 150 rails in captivity, how many members do we need to release to even have a hope of a successful introduction?" Pimm asked. "It could be that you could get away with one pregnant female. On the other hand, you might need thousands of rails for a successful introduction."

And then there was the question of where and how to put them out. They could all be released at the same time and same place. This would facilitate males and females finding each other, but would also risk the entire population being wiped out if a typhoon hit. They could be released here and there on Rota, which might enhance the chances for individual survival but reduce the chances for mating.

The question was how best to assure that a small population of rails grew. Any small population, Pimm knew, faces a greater risk of going extinct. What he was seeking in the release was a way to fend off those risks. The best way of doing that, Pimm thought, was to better understand why small populations tend to collapse.

Surprisingly, for very small populations it is not habitat loss, predation, pollution, or disease that is the biggest threat but simply bad luck.

Consider the smallest possible population—a single pair. If this is the population of some animal that lives for a year and during that time successfully produces a single pair to carry on, the question is how long the population will last. The answer is not very long. What if the environment is good and the population healthy? The answer is still not very long.

The reason is in the odds. There is a 50 percent chance of an offspring being one sex or the other. So, in the first year there is a fifty-fifty chance that the population will end up with either two males or two females. In either case the population goes extinct.

If the second generation does have a male and female there is still a 50 percent chance of their offspring being of the same sex the following year. In other words, this small population will always be on the brink of extinction because of demographics.

"There is nothing that you can do against demographic accident. It is a risk that befalls royalty. What they want are sons to take over, and you see time and time again how history has failed to oblige the kings and queens with the appropriate offspring at the appropriate times," Pimm said.

So small populations run the risk of demographic extinction even if everything else is perfect. Then there is the inevitable—death.

"You are going to die sometime, and suppose that there is a 30 percent chance each year," Pimm said. "Every individual in the population has a 0.3 chance of dying by this time next year. Quite clearly, there is a high probability that the two or three or four individuals in your population for quite independent and unrelated reasons are going to die before they breed again."

So the demographic risk and the steady tattoo of death, Pimm explained, "makes it certain that anytime you have very low populations, you have low survival times." In fact, when the rails were brought into captivity the first great crisis they faced was not the genetic bottleneck but these demographic accidents. But as the population grew, the risks of such bad luck diminished.

At the other extreme, Pimm said, is his "Ronald Reagan" model of species extinction. "Remember when the radio mike was left on, and Reagan joked about nuking the Soviet Union? Well, that's the other side. Not extinction based on demographic chance and independent reasons, but an environmental catastrophe where everybody dies for the same reason."

For something less than a nuclear holocaust, a typhoon, or a severe pollution incident, like the Exxon Valdez oil spill, the time to extinction for a population increases as the size of the population grows. If drawn on a graph with population size increasing along the horizontal axis and time to extinction increasing on the vertical, the demographic accidents are a concave line curving sharply upward and the time for environmental catastrophe is a convex line curving gently upward as populations get bigger and bigger.

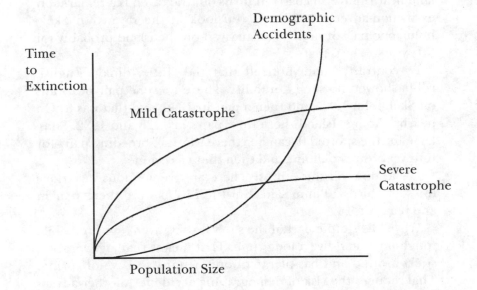

"What dooms small populations is demographic accidents. What dooms larger populations is environmental accidents. The practical question," Pimm said, "is where is that crossover point—one individual or 10,000 individuals?" That point—the intersection where the two curves met—was more than a theoretical construct. That point was the minimum number of rails that should be released on Rota.

To begin to answer that question, Pimm, in collaboration with Jared Diamond, a noted biologist at the University of California at Los Angeles, turned once again to data collected on islands. But the islands they looked at were subarctic pieces of rock half a world away from Guam in the waters of the North Atlantic. The data they scrutinized were bird censuses gathered diligently during this century by the British Trust for Ornithology and various amateur birdwatching groups. They were just raw numbers, but they were perfect grist for Pimm's theoretical mill. He considered the birds on each island as separate populations and then tried to determine which survived, which did not, and why.

He first considered the ravens of Skokholm, a cold chunk of land near the Welsh coast. "It turns out the raven is a remarkably extinction-adverse species. If you look at the ravens on Skokholm, one pair of ravens has survived on the island probably for fifty years. Why is that?"

By contrast, Pimm noticed that the "little, dinky" English robin showed no such durability. There was one pair of robins on Skokholm 1932, and then a pair in 1940, and that was it. On nearby Skomer Island there were pairs in 1928 and 1932. Similarly, for the spotted flycatcher there were six breeding pairs for four years on Skokholm, and then they were gone.

"What does all this tell us? There are tremendous variations for some birds," Pimm said. But the ravens were there year in and year out.

Why? The simple part of the story is that ravens are pretty big, tough birds and live a long time. That means that a generation sticks around and has plenty of opportunity to have offspring. That reduces the risk of demographic accident. Another advantage is that breeding generations will overlap, also reducing the risks.

By way of a supporting example, Pimm offered the story of another endangered bird on another island—the Bermuda petrel.

The bird once bred all over Bermuda. Now, however, it can manage to mate only on a tiny offshore island, because of rats on the main island.

Rats arrived on Bermuda about 300 years ago and since then the exact size of the breeding population has remained fixed at roughly thirty-five pairs. That number has been set by the number of rocky holes suitable for breeding on the little island. Without the holes, the nests are vulnerable to other predators.

A mere thirty-five pairs carry the petrel population each year. The remaining birds live offshore waiting for a chance to mate. And so the petrel must depend on a small breeding group to maintain its population. "Why on earth would that population last for 300 years?" Pimm asked.

"The reason almost certainly is that these birds live for about twenty years," he said. "So, even if there was some sort of accident that prevented the birds from breeding one year, they could always come back the next year and even if the entire breeding population was wiped out by a storm, they don't start breeding until they are six or seven years old, so all the other birds offshore could come back the next year and much to their surprise and probably delight, instead of finding all the breeding holes occupied, they'd find them empty. So that population, because of its overlapping generations and long lifetime, is really well adapted to putting up with demographic accidents."

Back on Skokholm Island, the ravens also carry on durably because they too live a long time and reproduce over several years. By contrast, the dinky robin lives little more than a year, and has to breed during that narrow window of opportunity. So, the robin runs all the demographic risks every year.

All this means is that species like the raven or Bermuda petrel—big, long-lived birds—are better at coping with the demographic risks of small populations than a little English robin with a brief life span.

So in theory a small population of ravens would have a better chance of surviving than a small population of robins. The bird counts from Skokholm prove that. But that still didn't answer the question of how many birds to release on Rota.

Pimm returned to his island data. He calculated the number of nesting pairs per island by species and grouped the thirty or forty species by whether they were large-bodied, like ravens, or small-bodied, like robins. "The raven hardly ever became extinct," Pimm said.

When populations on the islands were small, the "little guys" had a higher risk of vanishing. But as he looked at the population groups, Pimm observed something interesting. "There was a point at about seven pairs where the risk became just about equal," he said. By the time the small populations reached ten breeding pairs, the small-bodied birds fared as well as the larged-bodied ones.

And so, Pimm calculated that the point where the lines crossed was ten pairs. "What that says is that below ten pairs what knocks a population out is demographic accidents; above ten pairs for birds, it is environmental catastrophes that wipe them out."

What magical transformation happens at ten pairs? What

gives the small birds equal staying power with the bigger, tougher birds? The answer is resilience. While nature may have given the petrels and ravens many advantages over the robin and its tiny brethren, it balanced the ledger by endowing the small birds with the ability to reproduce quickly and massively.

Consider the story of the song thrushes in Great Britain. "You can see that something really pretty nasty happened in 1962," Pimm said. "It turns out that was a particularly severe winter. The song thrush population took about a 60 percent drop in numbers. But the population recovered those numbers in four years." The song thrush was pretty "resilient" in that it bounced back from an environmental catastrophe.

"Now," Pimm said, "let us suppose that these episodic catastrophes occur every five or ten years. If a species has low resilience, two catastrophes might do it—extinction."

Just as short-lived species are at a disadvantage when compared to long-lived ones, species with low resilience face a greater risk from environmental catastrophes than species, like the song thrush, that can bounce back quickly.

Ideally, a species that is long-lived and highly fecund will be the best survivor and be able to cope with the risks of demographics, death, and accident. "That's where the problem lies," Pimm said. For animals that are highly fecund tend to live only a short time and put a hell of a lot of effort into producing young. Animals that live a long time tend to produce fewer young. "Now, why does that have to be so? That is just the way things are. If you are small-bodied, you live a short time and produce a lot of young. If you are large-bodied, you tend to produce fewer young and live a long time."

The Guam rail happily fell somewhere in the middle. It was fecund, yet relatively large, and appeared to live at least a dozen years. All in all, it seemed to be pretty resilient. "There have been a lot of rail extinctions on Pacific islands, because they are running around on the ground," Pimm said. "But when you look at them as a whole, rails have generally been tremendously successful. They are everywhere, because of these characteristics."

So ten pairs of rails was the minimum that had to be established on Rota to avoid the ravages of demographic pressures and bad luck.

But how many birds had to be released to assure there were at least ten pairs, and how many more would be needed to assure that the rails could survive those more Reaganesque environmental catastrophes?

"Repeated introductions to different places at different times is the strategy that is going to minimize this unknown," Pimm said. But how many at a time? "It wasn't at all obvious to me what the number was. The easy answer was a lot of birds. But it is an expensive way of doing it and you risk losing all the birds."

The problem was there were only a limited number of Guam rails to release, because Susan Haig's analysis showed that a good number of individuals had to be kept in captivity to expand founder lines and preserve the population's genetic diversity.

Pimm had a few dozen birds to work with, at best. How could he simulate a large population and reduce risk? Pimm and Witteman went to the literature on American introductions of game birds—quail, pheasant, and such—to see how they had been managed and what success there had been.

These introduction records indicated not how long the populations lasted but simply whether the introduction was a success. The two biologists looked only at successful introductions and sorted them by the number of birds released. They found that the releases varied from six to 10,000 birds.

"We looked at the proportion of those introductions that succeeded and we noticed it was on the order of .04 or .05, one in twenty. Obviously, the more you introduced the more successful the introduction. But when you get up beyond 200 that proportion doesn't increase anymore. It levels out at 15 percent—one in seven. That says that even if you know, with hindsight, that this species is going to be successfully introduced, only one in seven times will it actually succeed."

The other piece of information gleaned from this exercise was that birds caught in the wild and released in some other habitat did much better than captive-bred birds, succeeding 50 percent of the time.

As they ranged over the data, Pimm saw that the ideal number seemed to be around eighty birds. "That left us with the situation that the minimum number we could release to get ten pairs was twenty and the optimal number was about eighty. So we settled for something in the middle—thirty. Why not forty? We only had so many birds available," he explained.

And so he had his number, thirty rails. Thirty rails at a pop would in all likelihood be enough to establish ten breeding pairs and that would be enough to beat the demographic odds. If they did periodic releases of thirty birds that should help cut the risk of environmental catastrophes. Of course, Pimm's work had also warned him that the odds of a successful reintroduction were probably less than fifty-fifty and perhaps as low in one in twenty.

It was a neat piece of work, and one Pimm believed would have relevance beyond classical ecology studies and far beyond just getting a handful of rails to breed on the remote island of Rota.

The "wild" was becoming nothing but a vast archipelago of isolated habitats dotting the globe, merely fragments saved from or yet unreached by mankind. It was time, Pimm argued, to think of untouched nature no longer as the rule, but the exception.

"There is almost nowhere on earth that we haven't trashed the environment; virtually all habitats have been altered," he said. "What ecologists do is go after the pristine believing that to be real. The reality is that the world is not pristine anymore. More to the point, it isn't going to be pristine in fifteen years. It is the extinctions that are going to be the reality.

"We are going to take the tropical forest, we are going to take all the important habitats on this planet, and we are going to destroy some of them and we are going to seriously fragment the rest of them. What the world is going to look like fifty years from now is little islands of native vegetation surrounded by highly modified areas. At that stage of the game, we are not going to be able to save nature on any grand scale, except in a very few special places."

In this race against destruction, the basic strategy has been to set aside as much land as possible in preserves. In the United States, a network of wildlife refuges have been created. In South America, swaps of rain forest in exchange for retiring national financial debt have been promoted by environmental groups and Western nations. But as the wilderness becomes more and more splintered, Pimm believes, these efforts alone will not be enough.

As Terborgh, Diamond, and Wilson had all warned, the smaller the fragments and the more isolated they are the greater the risks of extinction, and in 1991 the 5,000 refuges and parks that existed around the globe protected only 3 percent of the planet's land.

"We are going to have national parks and they are going to be surrounded by areas that are full of people, full of invasive plants and animals, areas that are ecological disasters," Pimm predicted. Such preserves will always be under assault from exotic species, disease, and human interference and suffer the risks run by small island populations.

"Fifty years from now people are going to be managing the animals and plants in those habitat fragments. We are doing that now in Hawaii and we are losing. If I was able to snap my fingers and get all the land I wanted in Hawaii, that wouldn't radically alter the problem.

"We still have the problem of managing invasive species, of managing very, very rare species. Look at Hawaii to anticipate the problems we are going to have in world parks a generation from now."

The wilderness will have to be supervised—using the tools of genetic analysis, captive breeding, population management, the frozen zoos, and a host of other disciplines. Individual animals may have to be transported from one nature fragment to another to maintain genetic diversity just as they are moved from zoo to zoo now in species survival plan programs. Biological controls, like Tom Seibert's moths, may have to be introduced from time to time to fend off invading plants. All the things being learned about managing islands may ultimately be used to manage what is left of nature.

"So what you are doing, you say, is training a generation of park managers," Pimm said. "Perhaps that's true. But what you are also doing is learning how to manage tremendously complicated natural systems.

"If we could manage communities, ecosystems, in the fragments we will have achieved an enormous amount. The science behind that is staggeringly complicated. But it is also enormously rewarding intellectually."

Pimm couldn't wait to get started. He had his number, his research plan, his funding, his rails. Now, all he needed were the necessary bureaucratic approvals. But for some reason they weren't coming, and it was getting perilously close to the time he had assured his funding sources that rails were going to be released on Rota.

Chapter

18

January 1989

The gathering had the conviviality of a cocktail party, the expectant air of a movie premiere. Stuart Pimm's friends and colleagues sipped wine and milled around the large living room of his woodsy, hillside home, chatting and keeping an eye on the television set tuned to the local Public Broadcasting System channel. "I had," Pimm said, "invited my friends over to watch my fifteen minutes in the sun." At eight o'clock the theme music for "Discover: The World of Science" began. The moment had arrived—celebrity.

The PBS show had prepared a segment on the rail and the brown tree snake. They had visited Front Royal and filmed Scott Derrickson and Susan Haig. They had traveled with Pimm and eight rails back to Guam. They had shot "native" dances, which, if the truth be told, were a bogus exercise staged for Japanese tourists, and had captured the beautiful vistas of Rota.

And there on the screen was Stuart Pimm, ever the consummate grantsman, wearing an International Council for Bird Preservation cap and a World Wildlife International T-shirt. "Every time I was on camera, I tried to do this," Pimm quipped, expanding his chest, an effort to give his sponsor some extra billing. Yes, there was a bit of everything in the documentary, everything except the release of the rails on Rota. The project was now six months behind schedule, and there was no prospect of an imminent release. Birds had now been shipped back to Guam—in the passenger compartment of a Continental Airlines jet—but no rails had yet been set free.

Still, it was publicity, and that was something. "They did a good job," Pimm said, "and I thought: 'This is it, fame. Everybody will be queuing up to give money to the Guam rail project.' But there was no effect whatsoever. I had great hopes that this would generate a lot of support for the program because by this stage we were really running low on money . . . but it didn't. It had almost no effect."

By early 1989, Pimm and Witteman were moving into year two of what Witteman had dubbed "the great wait." Pimm had purchased plane tickets, soon after Witteman had arrived in Knoxville, assuming that it would only be a matter of a few months before they left for Guam and Rota. Wrong. The project needed federal approvals, approvals that seemed to be taking forever to come.

"There was an expectation on the part of the DAWR and Bob [Beck] in particular that the permitting process was going to be easy . . . but it had to go through the Fish and Wildlife Service bureaucracy and then appear in the *Federal Register*," Pimm said. "Any reasonable expectation was that the permitting process was going to take a long time. It was naive to think it would be smooth."

Pimm was getting nervous. "We'd had money from these guys for almost a year now, but we hadn't done anything with it." That wasn't quite true. Part of it had gone to pay Witteman, until Pimm was able to get him a teaching position at the University of Tennessee, and part of it had gone to purchase a computer and a truck for the work on Rota.

When Pimm arrived in Micronesia in August 1988 with the PBS film crew, his anxiety turned to gloom. "It was at that point that it became clear to me that nobody on Guam had an idea how long it was going to take to get a permit," he said.

"It was remarkable we had gotten the money we had and I was very, very afraid that we were going to have to ask for another ten or twenty thousand in order to do what we had failed to do before," he said. "That wasn't made any easier when we got to Rota."

The project's pickup truck had been delivered to the island two months earlier—registered and insured in Pimm's name. "So, I naturally asked for the vehicle to be there when we arrived at the airport," he said. But when he landed on Rota there was no truck. A taxi ride brought him to the CNMI Department of Natural Resources, where the truck was supposed to be garaged. Still no truck. A few phone calls later the truck was located and brought to the government office. It had a dent in it and, on an island with only six miles of paved road, there were 8,200 miles on the odometer. It had apparently been driven about 100 miles a day since it was delivered in June. It shouldn't have been driven at all.

Pimm was mortified. "Oh, God, I'd have to say: 'We've spent $40,000 and haven't released a rail. We can't give you your truck back because it has been driven 10,000 miles, but we can give you this $2,000 worth of computer equipment,' and I thought, 'There goes my chance of getting a grant to do conservation biology, ever again.' "

Back on Guam, however, Bob Beck thought everything was going along fine, if a bit slowly. "Unfortunately this is a very long and involved process," he said. "Getting this designation requires *Federal Register* notification; it requires publication, and furthermore it requires an environmental impact statement on the effects of the release to demonstrate that this whole process isn't going to have an adverse impact on Rota. So this is a very time-consuming, bureaucratic process."

He explained that there is a prohibition in the United States against introducing species outside their native range, although, of course, it is done all the time by people carrying plants and animals to places they don't belong. Hawaii is overrun with banna poka from South America and gorse from Scotland, both brought to the islands by farmers as ground cover. But the DAWR was a government agency and had to play by the rules, and Rota was not the natural home of the Guam rail. "We had to get a special dispensation from the feds to allow us to put the bird on Rota," Beck said.

In addition, the CNMI and Rota's mayor, Prudencio Manglona, were concerned about putting an endangered species on Rota and what that would mean in terms of tying up development on the island. "So to get around that, we applied for 'experimental population status'—which is in the Endangered Species Act," Beck said. "This allows some of the prohibitions of the act to be relaxed a bit, so that things aren't quite as rigid, and this satisfied the CNMI."

The DAWR's environmental assessment had to show that the rails wouldn't hurt the other animals or native plants on Rota. That required knowing the native flora and fauna on this remote Pacific island and that forced the DAWR to do the very first surveys of insects and reptiles of the host island. All this took time.

Then the release proposal was drafted by the DAWR, edited by Fish and Wildlife Service officials in Honolulu, and passed on to the regional office in Portland, and from there it went to Washington, D.C. It made the trip from office to office several times before everyone was satisfied.

"Nothing went wrong as far as we can tell," Beck said. "We had no significant detours. It was just bureaucratic inertia. It just took a long while to have everybody review the project and sign off. It didn't appear to be any significant opposition to the whole thing.

"At one point I had written a memo to everybody. I realized it wasn't going to occur when Stuart thought it was, and he didn't like that because he had a grant from January '88 to January '89 and of course a release was supposed to occur during that time period. So the thought that it wasn't going to occur during that time was really upsetting to him. I can't imagine why it should be upsetting to the granting agency. These are folks who are used to conservation projects. Besides, it was out of our hands anyway. The feds were calling the cards on that one. We couldn't make a move without them."

Winter slipped into spring and spring into summer; finally Pimm could stand it no longer. The first notice of the rail release was placed in the *Federal Register* by the Fish and Wildlife Service in July 1989.

At best, it would still be three months before the public comment period would be closed and final approval granted for the federal permit to release the birds. But Pimm and Witteman could wait no longer. They headed for Rota.

"At that stage I decided it was time to move out there," Pimm said. Immediately upon arriving Witteman informed his doctoral advisor that the island's power supply was so erratic that the project's personal computer could not safely run on it.

Pimm flew to Guam and bought a huge rechargeable battery for $1,000 on his American Express card, not quite knowing where the money for the unexpected expense would come from. They had already bought sheets, pots, plates, and other sundries and Witteman had set up shop in a two-bedroom apartment at the Senior Apartments, a modern stucco building on the far side of Songsong, overlooking sparkling Sonsanjaya Bay.

Witteman had air conditioning and a kitchenette with an electric stove and a large refrigerator. Pimm had hoped that they would be able to find cheap housing on a backwater island, like Rota, but, as he put it, "There were only two kinds of people who lived on Rota: people who *really* lived on Rota and people who came under contract to the Commonwealth of the Northern Marianas Islands to teach."

The CNMI contract provided a housing stipend of "up to" $650 a month, and so every "apartment," from a tin shack with a dirt floor to Senior, which was the top of the line, cost $650. It was a budget buster, but they had no choice. Witteman figured that if everything went well, he could make the grant money last until the end of the year.

"We got Greg settled and then I shook hands and I said, 'Good luck old chap,' and went off," Pimm said. "And there was Greg, poor chap, sitting on Rota alone, waiting for the approval of the release."

Culture shock. Greg Witteman had grown up in Blue Jay, California, a small town near Big Bear Lake in the San Bernardino Mountains, but a small mountain town and a small island are very different places.

"Rota is a real, real small town, but a different sort of small town than Blue Jay, where you feel like an American. Going to Rota was like going to a different world. It was sort of like going down to Baja. The same sort of 'Tomorrow,' or 'My cousin will do it for you,' or 'It's really not that important.' You know what Hawaiian time is? Rota is Hawaiian time to the max. The attitude is sort of 'It will get done eventually; somebody will do it.' "

The first week he was there, the island was blacked out for almost two days because the key to the diesel fuel intake was misplaced and the generator ran out of fuel. Witteman was baffled by the constant digging up of Songsong's streets by public works crews. It turned out that the water mains were so old they continually sprang leaks, which would be repaired by wrapping old inner tubes around the pipes. This, of course, assured that sooner or later the crews would have to come back and dig them up again. He was surprised by the constant traffic, in a town that couldn't be more than half a mile wide. He found, however, that it was so hot, the roads so dusty, and sidewalks so nonexistent that a stroll across town would reduce a pedestrian to a sweaty, sodden mass.

It is difficult to ascertain what the Rotanese thought of Witteman, who trooped around in torn T-shirts, baggy work pants, sandals, and a huge leather sombrero, with two days worth of stubble on his chin. His aura was more one of a desperado or smuggler than a biologist.

Aside from public, make-work projects and driving around in government vehicles, there didn't seem to be very much "business" going on on Rota. But with a tropical climate, generous government aid, and a little subsistence farming it was easy to get by. "For a lot of people there isn't anything they really have to do," Witteman said. "It's sort of a quiet place. Probably nobody has ever died of an ulcer on Rota."

But Witteman did have work to do. He began building a radio-tracking platform for the back of the pickup truck and preparing the release site up on the plateau, or sabana, as islanders called it. A set of cages had already been constructed in a grass field by DAWR, and Witteman started scouting the surrounding sabana and rain forests to learn the terrain and see what problems the birds might encounter.

Rota's sabana was a completely different world from the shoreline woods below or even the forests sitting atop Guam's limestone bluffs. Because Rota's cliffs are more than twice as high as those on Guam, the sabana is perpetually topped by a cap of clouds.

"It is a nice flat top with a high elevation," Witteman explained. "It hits the clouds and makes it rain. It is foggy and wet all the time. . . . During one typhoon there were twenty-three inches of rain in a twenty-four-hour period up on the sabana. At the airport they only counted six." Scott Derrickson described it as being "as close to a cloud forest as you will find in that part of the Pacific."

There is almost always a low dome of white and gray clouds above and the sun is no more than a pale disk, rippled by the wind-driven mists. The light falls in filmy sheets that seem so thick, so tangible, they could fill a bucket. When the clouds break and the sun shines, it transforms the sabana's expansive sawgrass fields, little farm plots, and thick forests from a deep, woodsy green to glittering emerald. It all sparkles because everything is coated with a perpetual dew. Then the clouds come again and, like a switch being turned off, the trees and grass stop shining.

The Pacific Ocean, save at the cliff line, does not intrude on the sabana's world. The surf's roar is supplanted by the wind, and the sea cannot be seen or smelled or felt.

The forests, little bothered by humans and a little more out of the way of Pacific typhoons than those on Guam, have grown lavishly. Wading into the woods, a hiker is engulfed by the leaves and branches, and he disappears into layer after layer of moist, organic green. There is no frame of reference, no perspective, for there is always a dense wall of vegetation a foot from one's nose. Ripe breadfruit, the size of bowling balls, hang from the trees above. It is so still that the buzz of bees, the flap of the wing of an unseen bird can be heard.

This is where the rails would be released—if the federal government ever approved. And this is the land Witteman set out to explore.

Sometimes Witteman worked methodically, conducting bird surveys along transects, setting lines of rat traps to measure population densities, or mapping geographical features to find the places where they risked losing birds. "The island is basically a raised limestone plateau that has a lot of holes in it, and old coral fingers that are two or three feet high and extremely sharp," he explained.

Then there were days he would just explore, and his wanderings weren't limited to the sabana. Witteman hiked through the seaside forests checking monitor populations and climbed down the rocky promontory of Saguagahga Point to an extensive seabird colony on the northern end of the island. "I spent a night in a cave down at the seabird colony to see which birds were calling. It was a nice dry cave," he said. "I just wanted to look at the various habitats on the island and see how very different they were and if there was sort of quantitative and qualitative difference we might be able to exploit in the future."

Most of his days, however, were spent up on the sabana—and sometimes his nights too. "I'd go up there, pull off over toward the cages, and walk around all day. Sometimes I'd take a compass. I was really supposed to take a compass all the time, but one time I forgot and I got lost in the rain forest. I was sort of following the sun and it got cloudy. I figured I would turn this way toward the road, but it was pretty dense and cloudy and it rained and it got foggy and there was nothing I could do," he said.

Witteman decided to wait it out. "I felt pretty silly. I was searching a new area of the rain forest. I figured the birds wouldn't go into it because it was so dense, but I thought, 'I'd better look in and see what's actually in there.' I found out. There was lots of 'wait-a-bit,' this very tough stringy crawly vine. It is a woody vine with few leaves but the entire surface is covered with back-pointing claws that are hooked on the end like fishhooks so it gets on your skin and on your clothes and makes you 'wait a bit.' It's really painful. I still have a scar from wait-a-bit actually ripping across my arm.

"I thought I could go up a tree and see my truck right around sunset. But it was no good. I got up about halfway in this breadfruit tree and found this nice fat branch and slept there a couple of hours. There were lots of ants all over the ground. . . . I had brought a flashlight, not a compass. That was pretty silly. The flashlight didn't do me any good because I could only see about ten feet in front of me.

"I had my parka on and my backpack full of things, water, a granola bar. Actually it wasn't a granola bar, it was something with Japanese letters on it that was edible, some sort of chewy substance, and I had this Gatorade substance called Pocari

Sweat. They don't have to label things on Rota so you just have to guess what's in it."

The next morning the sun came up, the fog lifted, and Witteman saw his truck—about six hundred feet from his breadfruit tree.

Chapter

19

August 1989

A strange, troubled cry woke Ernest and Yvonne Matson at dawn. Their ten-month-old son, Skyler, was bawling, but it wasn't the "I'm wet" or "I'm hungry" or "Why isn't somebody here?" cry they had come to know. Matson, a marine biologist at the University of Guam, walked through his apartment in the rural, seaside village of Ypan and into Skyler's room. There he found the boy wrestling with a five-foot-long brown tree snake.

"He was standing in the crib doing his damnedest to get the snake off him," Matson said. But the snake had wrapped its tail around the boy's neck and the rest of its body was coiled around the boy's middle. Then, Matson saw the large patch of blood on the sheet. He pulled the snake from Skyler, killed it, and then examined his son. The baby's left hand was covered with bites and salivalike venom. "The snake couldn't get Skyler's hand into his mouth, but he was trying," Matson said. The arm was already beginning to swell.

Along the coast road and through the hills, the Matsons quickly drove to Guam Memorial Hospital. The emergency room doctor told the couple not to worry. He explained to the Matsons that the snake was not poisonous. Perhaps the doctor believed that, but people on the island with small children knew better.

Since the first reported incident back in 1987, twelve small children had been attacked in their cribs by the brown tree snake. Just that March, Air Force Staff Sergeant Donald Hilton and his wife, Carmelita, had taken their five-month-old daughter, Susan, to the U.S. Naval Hospital after she was attacked by a snake. The baby appeared to be okay, but right there in the emergency room, Hilton later told the *Pacific Daily News*, "it hit her all at once. . . . She quit breathing. She turned fuzzy and slumped over."

According to medical records, her heart rate jumped from 140 to 240 beats per minute, then slipped to 70, and she went into respiratory arrest. The infant was placed on a ventilator with oxygen, and suction removed "copious" secretions from her airways. It was four days before Susan Hilton was discharged from the hospital.

Still, the Matsons were assured by the hospital medical staff that everything was okay. Matson wanted his son admitted. "I almost got into a fist fight with this doctor who was trying to tell me the snake wasn't poisonous." But the doctor prevailed. The bites were cleaned and dressed and Skyler went home with his parents.

By 10 A.M., however, Skyler's arm had swelled so that it "looked like Popeye's," and the boy was "droopy and lethargic." Within an hour Skyler couldn't even stand up or talk. "He was limp, like spaghetti . . . a puppet head," Matson said. The family drove back to the hospital. They were told, "Don't worry, he'll be okay." But with the story of Susan Hilton lingering in his mind, Matson would not be turned away. "I was yelling and screaming. My wife, who was seven months pregnant at the time, was in tears, borderline hysterical," he said. Matson demanded that his son be admitted and what's more wanted the boy in intensive care with a cardiac monitor. He got his wish. The boy was placed in pediatric intensive care with a monitor and was also given an intravenous glucose solution.

The Matsons spent the night at the hospital. Skyler lay there senseless. The monitor showed that his heartbeat was irregular and slightly elevated. But, as the hours passed, he appeared to stabilize, and by early the next morning the brown tree snake venom's hold began to weaken. "He was coming around. It was like he had a hangover," Matson said. By 4 P.M.—about thirty-four hours after the attack—he was once again normal.

Yvonne Matson took to covering drains at night with pots, stuffing newspaper between pipes, and sealing cracks. Her husband, however, was dubious about these efforts. "There is nothing you can do to keep the snakes out. . . . You can't snake-proof a house on Guam." The slippery rural roads, he said, remained more a menace than the snake. "What bothered me the most was the hospital people telling me the snake was not poisonous."

It appeared that the brown tree snake was beginning to write a new chapter in its sojourn on Guam. The reptile was now attacking babies, and island public health officials and herpetologists were at a loss to explain why. "There are hardly any incidences of snakes attacking sleeping people," Fritts said. He added that there were cases of a species of krait, a venomous, nocturnal snake from Southeast Asia, attacking sleepers, but that was the only one he knew of in the world. Certainly the brown tree snake in its natural range was never known to do such a thing.

Fritts examined seven of the serious bite cases and said that in five, the parents stated that the snake was constricting as well as biting the child. "Snakes don't constrict you when they are in a defensive mode. A snake, when he grabs a rat to eat, will throw some coils to hold it while he is chewing, controlling the prey," he explained. In addition, five out of the seven had multiple bites, another indication that the animal was probably trying to chew on the baby.

"All of these children were in their beds at night. They weren't threatening the snake, or clubbing it, or playing with it. . . . Some of these kids had three or four fingers chewed up. So they are feeding bites," Fritts said. "Feeding bites may involve injecting more venom. But that's a lot of conjecture on my part."

What they were seeing was feeding behavior from the larger brown tree snakes on the island. Most of these attacks involved snakes more than four feet long. The snakes in both the Hilton and Matson cases were five feet long. That appeared to explain what was happening, but it didn't explain why it was happening.

Reviewing the cases in 1990, Dr. Robert Haddock of the Guam

Division of Public Health said: "The bizarre thing is that it seems to seek out babies. . . . This seems to be a new phenomenon; it isn't something we were aware of eight years ago. . . . It is trying to adapt to eat humans. It hasn't, but it is certainly trying."

What troubled Haddock most was the way the snake seemed to find the children. "We had one case where the infant was sleeping between the mother and father, and somehow the snake distinguished between the adults, and they found it chewing on one of the baby's fingers."

Fritts's analysis of the problem began with the population profile of the brown tree snake researchers had compiled. "The only place you can find big snakes is in the urban areas because that's where the chickens and pigeons and black rats are. The snakes living way out in the boondocks without that influx of large prey have much less chance of growing that big.

"Once they are in the urban areas and get into homes, they see these babies. They think, 'Well, this is the smallest thing here, maybe a very big rat . . . I'll try it.' " Just as the snake had adapted to hunting on the ground when necessary, it appeared the animal was now trying to make yet another adjustment.

But what was drawing them to the babies? Perhaps it was something about the way the infants smelled. One hypothesis espoused was the "milk-breath theory." Babies drink milk. Snakes like milk because of the fats or lipids. "We know snakes like lipids. You can get snakes to drink milk. Nicotine sulfate [snake poisons] were offered in milk. So maybe it's the milk drool of human babies that is attracting them," Fritts speculated.

There were, however, other possibilities. "A female dog or its puppies are much more likely to be attacked the night they are born than three nights later," Fritts said. "It is often at the time when a bitch has pups that you find that a snake has gone out and killed or constricted one or more puppies, whether or not it's been able to eat them." This predatory skill made sense to Fritts. "It is a lot easier to catch and eat eight micelets than two adult mice," he explained.

Julie Savidge had interviewed a pigeon fancier years before who had complained that the snake was always to be found when his pigeons laid their clutches. Clearly, "milk breath" was not involved here. Perhaps it was some hormonal odor that was enticing the snake. What link was there between pigeon eggs, newborn puppies, and sleeping babies? Tom Fritts was on the case. There were already two dozen snakes in a lab in the Rocky Mountains being probed for the answers to these questions. As abhorrent as the idea of snakes stalking babies was, maybe this would be just the break they needed to find a chink in the brown tree snake's biological armor.

In addition, Fritts had arranged for fresh troops on the island. In October, Mike McCoid arrived on Guam as Julie Savidge's replacement at the DAWR. Fritts had been cochairman of McCoid's master's thesis committee at San Diego State and had first alerted his protégé to the Guam job. At the time, McCoid was in a Ph.D. program at Texas A & M, working on the taxonomy of an obscure group of Central American fish and not caring much for it. For him, Guam, as it had been for Beck, seemed an ideal refuge.

A six-footer, with curly brown hair, a brush of a mustache, and windowpane-sized glasses, McCoid was an army brat who easily bridged the gap between tropical island and military installation. Of course, he had been hoping for more tropical island and less military installation on Guam.

"I was really disappointed. . . . I expected something more pristine and undisturbed. This was like being in Dubuque. . . . Any place in the middle of the Pacific, in the middle of nowhere, that has Pizza Huts is not a tropical paradise. On the other hand, you don't lose very many creature comforts. It is all there, the supermarket, the mall—and Domino's delivers."

One of the things McCoid started to do was to collect data on the snake bites. He and Fritts drafted a "snake bite incident form," which he distributed to all hospitals and health clinics on the island.

Fritts was concerned about the attacks not only because of the threat they posed but because he feared a public clamor might goad politicans into committing a lot of funds for research into a brown tree snake antidote.

"It's not like we need an antivenin. We just may need a better emergency room protocol," Fritts said. "The symptoms we have seen in the serious bites can be treated by any attentive emergency room physician. It's the inattentive one I'm worried about. The one who is not watching the child, who sends him home saying, 'This snake is nonpoisonous; put your child to bed and he will be fine in the morning,' and he is put to bed and he goes into respiratory arrest with no one watching him. He may be dead in the morning."

Like Savidge before him, McCoid continued to collect and dissect snakes. By 1990, the collective efforts of Savidge, Fritts, and McCoid had resulted in more than 2,200 necropsies on brown tree snakes. But McCoid added a new measurement to the data—fat weights. He noted precisely how much fat there was on each animal.

Everyone knew that brown tree snakes were mean. McCoid was finding that they were also very, very lean. The snakes were stressed. Perhaps the principles of population biology were finally kicking in. As the density increased and the prey base narrowed, maybe food was getting harder to find. McCoid was finding very little if any fat on the animals he caught. But he still wasn't certain what that meant. For example, McCoid had opened female snakes, with no fat reserves at all, that were ready to drop clutches. Would such eggs be less viable? Would the female just die after the effort of reproducing?

Three nights after arriving on Guam, McCoid went snake hunting with Gordon Rodda out on Orote Point near the Naval Station. "Gordon and I were walking side by side, when he stopped in front of a tangled mat of vegetation and said, 'You are within three feet of a brown tree snake right now.' I stood there and looked, and I could not see it. Finally, Gordon had to go up and physically grab the snake. But I couldn't see it at all," McCoid said. "It took a while to get the right search image. . . . You can understand why some people have lived here for years and never seen them."

The following week, McCoid was awakened by something in his bedroom. He flicked on his night table lamp. There staring him in the face was a brown tree snake. There was no problem seeing this one; he was sitting coiled on McCoid's bed. There was a gecko on the wall behind him. "I must have moved or flinched. . . . The snake—it was a little one—shot out and bit me right in the cheek. If I wasn't a herpetologist, I'd have been on the next plane out of here," he said.

The attacks on babies, although not frequent, continued, with five more infants being bitten by early 1991. On August 8, 1990, almost a year to the day after Skyler Matson was attacked, Joyce Kaneshiro was preparing breakfast in her home in Agana Heights. Her three-year-old, Christina, walked into the kitchen but wouldn't respond to her mother's morning banter. "She seemed stunned," Kaneshiro said. Then her four-week-old baby, Carmarin, began crying. "Let's go give Carmarin her bottle," Kaneshiro said to Christina.

They entered the master bedroom—the children still slept in their parents' room—and Christina began to say, "Mommy, there is something on Carmarin," when Joyce saw the brown tree snake wrapped around the baby's neck. She screamed. Her husband, Derrick, leaped out of bed. He pulled the animal from his daughter and killed it.

There was blood all over the baby's arm. They called an ambulance and were taken to their family medical clinic, where the wounds were cleaned and the baby given a tetanus shot. Carmarin appeared to be fine but the doctor suggested she remain for observation.

Six hours later, they were getting ready to release Carmarin when, Joyce said, "she started to cry, and her cry sounded as if she was being choked. One of the nurses advised me to get her bottle and feed her. But Carmarin just drooled. She couldn't swallow. The doctor saw she was having trouble breathing and transported her to Guam Memorial Hospital for oxygen."

By the time Carmarin arrived, the emergency room staff was no longer casual about snake bites. "Her heartbeat was abnormal and she was having trouble breathing. . . . They hooked her up to oxygen and an IV and they were injecting all these needles into her. . . . It was terrible. I panicked all over again."

The infant was admitted to intensive care pediatrics. One of the doctors told the Kaneshiros that the bite didn't look too bad. A couple of the cases that had come in, he told them, had looked like "ground beef." A day and a half passed before the baby was released from the hospital. "It was awful," Joyce said. "I almost lost my baby to that snake."

20

November 1989

Could things be any worse? Witteman was broke. Rent and utilities had just been more than his budget would bear. He couldn't pay for his flat at the Senior Apartments, and he had moved out. The only lodgings he could find on Rota were in a cinder block house that island agricultural officials had used to store pesticides for a melon fly eradication project. The agricultural agents had been rather careless, and the building was heavily contaminated with Malathion.

"I went in and hosed the back room, which was about ten feet square, chipped off all the paint, and made gaskets for the door so that the contaminated portion wouldn't contaminate me," Witteman said. "I moved all my equipment in there and crashed there at night." He had barely gotten all this gear in when a typhoon hit, so there Witteman sat in the Malathion house, a storm swirling around him, with no air conditioning, no money, and no idea how he was going to complete his project. All he had was the return plane ticket to Tennessee. "I assumed that there was money somewhere but that I just couldn't get my hands on it."

It was a rash assumption. Back in Knoxville, Pimm was drafting an artful letter to his funding sources describing all the progress they had made, the great prospects for the rail release, and, oh yes, the dire need for some more cash. "We were out of money. My worst nightmare had happened," he said. At this point, Scott Derrickson stepped in and passed the hat around among the zoos breeding rails. "He was fantastically helpful. Scott gathered $50 to $200 from each of the zoos," Pimm said. That $5,000 kept the program afloat, and Wildlife Conservation International, much to Pimm's relief, indicated that they, too, would provide additional funds, but only $13,500, less than half of what had been asked for.

Still, the project was saved, at least for the moment, and the first week in December, Witteman moved back into the Senior Apartments. "By that time it looked like the Hilton, the Rota Hilton," he said.

Nevertheless, this brush with bankruptcy left Pimm nervous. "We were down to the stage where we were barely able to survive. . . . I was very concerned that when the release took place, it should generate great enthusiasm for the project, and so I spent a lot of time trying to engineer the release in such a way that it would attract the right kind of attention," he explained.

The big event was now scheduled for February 1, 1990. Pimm managed to convince S. Dillon Ripley, who would be on his way to India, to stop on Rota and ceremonially release the first rail. He spent time trying to drum up media interest. He pitched the story to the networks without success but did manage to convince CNN that it would be worth a news spot. This was, he kept telling foundations and news organizations, the first time in history that an American animal completely extinct in the wild would be returned to nature. "It was going to be a historic event," Pimm maintained.

It was amid these preparations that Pimm got a call, in the middle of December, from someone at Wildlife Conservation International asking whether he had seen the report in the *New York Times* about the rail release. What rail release? Pimm wondered. There wasn't, he fumed, supposed to be a rail release.

But on the other side of the world, Beck had thought differently. "We knew," he explained, "that this release was going to have a lot of press coverage. We knew that this release was going to have the high-level feds there. We knew Dillon Ripley was going to be there. We knew that there were going to be bigwigs around, and we wanted to be sure it wasn't going to be a total disaster." And so, on December 9, Beck and Witteman released six rails on the sabana. The birds scuttled off into the thick sword grass, without incident. Everything seemed okay. "It was practical politics," Beck explained. "The worst thing that could have happened to us was to release all these birds and have them die immediately because of some ridiculous reason."

It was a clear and windy morning on the sabana, with its ever-present cloud ceiling, when a line of trucks, jeeps, and cars wended its way up the steep, rutted road, that first day in February. Pimm was there. Derrickson was there. Muna was there. Beck and Witteman were there. So were representatives of the Fish and Wildlife Service, the CNMI, and the DAWR. And S. Dillon Ripley, Mrs. Ripley, and his personal photographer were there.

Muna had become an expert at rail husbandry and was now in charge of handling the birds for release. He would pull a rail from its holding cage (the birds had been given identification leg bands and placed up on the sabana several days before to acclimate them to the area), bring it into the cab of a truck, and slip on the little radio transmitter backpack. The bird was then ready for release.

With the exception of Witteman, Muna, the Ripleys, and the photographer, the rest of the party—about a dozen people—stood some fifty yards away from the release site and watched through field glasses and telephoto lenses.

Muna handed the bird to Ripley. Ripley stooped down and let the rail scurry through the sword-grass field and into a stand of trees. It took about an hour to release the first six birds. The tall, patrician Ripley strode back toward the waiting onlookers. " 'Tis done, 'tis done," he said with a dramatic, stentorian flourish. "The transaction's done. . . . *Henry V*, I think." The party climbed back into their vehicles and lumbered down the road to a fiesta—complete with a whole roast pig—held in Ripley's honor by Mayor Manglona.

Although a reporter had traveled all the way from Philadelphia to cover the release, the CNN crew never made it. It turned out that there was big news on Guam that very day.

About the time that Dillon Ripley was preparing to release rails on Rota, Ricky Bordallo, the former governor of Guam, the man who had called Julie Savidge back in 1983 to tell her not to worry about the brown tree snake and look at pesticides, was scheduled to get on an airplane and report to federal prison in California.

The sixty-two-year-old Bordallo had been found guilty and sentenced to four years in jail for witness tampering and obstruction of justice in an investigation of $600,000 in illegal campaign contributions from Japanese businessmen who were building the big, shining hotels along Tumon Bay and knocking down the forests to make way for golf courses.

Something as rude and traumatic as the Bordallo case had never happened before on Guam. It wasn't part of the way life was conducted on the island. Bordallo had been the scion of a politically powerful family, the owner of Ricky's Toyota Dealership, and a long-established politician in his own right, one of Guam's good old boys.

For years, the Bordallos had been Guam's first family. When his daughter Debbie flew to London to participate in a beauty pageant, the papers enthusiastically reported the event, and each Christmas there was a full-page picture in the paper of the Bordallos around the Christmas tree, wishing the people of Guam season's geetings.

Now it was all over. Getting on that plane to go to the mainland and sit in prison was more than Bordallo could bear. So instead of going to the airport, he drove down to Marine Drive and the statue of Chief Quipuha. There he wrapped himself in a Guamanian flag, took out a .38-caliber pistol, pointed it at his right temple, and blew out his brains.

Too bad for the rails. This was perhaps the best ninety-second prime-time news story ever to come out of Guam. The CNN crew never got on the plane to Rota. "It was terribly inopportune," Pimm lamented. "If only he could have done it a few days earlier."

Over the next week, ten more birds were set free—for a total of twenty-two, counting the December release. Not exactly the thirty birds Pimm's calculations had suggested, because the DAWR was forced to hold back several birds for health reasons. By then all the dignitaries and senior scientists were gone, and it was time for Witteman's real work to begin.

The goal of the radio tracking was to see how the birds established territories, if males and females found each other and mated, and in general to construct a picture of the natural life of the rails. Over time, as more and more birds were released, Witteman would be able to compile detailed information on how the rails adapted to their new home. As Gary Wiles had observed, radio tracking was a time-consuming exercise, but this was precisely why Witteman had come to Rota.

Each morning, Witteman and Stan Taisican, a CNMI wildlife technician, drove up to the sabana at about five and sat, very still and quiet, at the intersection of the road and a jeep trail, looking for rails with a 60-mm telescope.

"We'd do that in the morning—just sit there in the road—until we could see the sun in the sky, because after that Filipino field hands would come up to work people's fields. So there was traffic on the road," Witteman said.

"After we'd finished the visuals, we'd radio track birds we hadn't seen," Witteman explained. "We'd go to different locations—there were checkpoints we'd set up every two-tenths of a mile—from the edge to the center of the sabana. Sometimes I'd go in the truck, sometimes I'd just walk it."

Witteman carried in his hand something that looked like a TV antenna from some suburban rooftop and a small, portable radio receiver draped about his neck. "Each one of the birds had a different frequency and each radio beeps about every second. So, you'd go there with a highly directional antenna and you'd do a frequency of one bird and find out what direction it was coming from. I'd scan and write down frequencies that were beeping.

"After I'd written them down, I'd turn off the scanner and turn it to manual and toggle it up to a frequency and using a compass try to locate each bird. The rain forest blocks out the signal pretty well after 300 meters or so, so I'd have to walk in there or walk around the other side of the woods."

This was how it was supposed to work, ideally. Reality turned out to be more frustrating. Even before the official Ripley release, Witteman had completely lost contact with two of the six birds freed in December. A third December bird had died and a fourth was giving off an odd signal. The signal was puzzling because it appeared to be stationary. Every day, Witteman had been able to pick up the signal in the same place, but as he moved toward it, it vanished and then seemed to be coming from somewhere else in the woods.

"I couldn't figure out what was going on. But one day I was standing at the edge of a forty-foot hole. There are these natural shafts where the water has eroded the limestone, you've got to be careful or you can walk right into one, and I pointed the antenna into the hole and that's where it was coming from," Witteman said. The signal had been elusive because it was rising out of the hole and bouncing off the trees.

It was New Year's Day, but Witteman was determined to find out what had happened. He returned to town and borrowed a lineman's belt and some guidelines from the telephone company, a pulley from the local dive shop, a strong rope from a fisherman, and an old, broad-brimmed fire hat from the Rota Fire Department.

Then he returned to the pit with Taisican. Witteman's plan was for one of them to be lowered into the hole to search for the radio. Upon further consideration, Witteman thought Stan—who weighed perhaps 130 pounds and stood five feet seven inches—ought to go. "I explained that I could always pull him up."

A dubious Taisican, wearing the lineman's belt and the fire hat, was lowered into the hole. Forty feet down and about twenty feet wide, the bottom of the pit was thick with brush. Taisican found a tree stump and a deer skull and eventually the radio, but he never caught sight of a rail.

When it was time to come up, however, the crumbling limestone walls wouldn't permit it. Taisican could not climb and Witteman could not pull without causing a minor limestone avalanche. They tried for several hours. They divided their time between concocting stratagems for getting Taisican out and smoking cigarettes and chewing on betel nut, whose red juice provides a gentle, narcotic buzz. Finally, Witteman used a walkie-talkie to radio DAWR headquarters. He was informed that DAWR personnel could come out tomorrow with a block and tackle and get Taisican out. By now, it was growing dark.

"It's okay, Greg. I'm going to wait. Just toss me down some more cigarettes and betel nut," Witteman remembers Taisican saying.

"Stan, we don't have any more betel nut or cigarettes," he replied. "Then I heard this thrashing and rocks tumbling and slowly Stan, shaking and covered with limestone dust, climbed out of the pit," Witteman said.

The birds in the later releases fared little better. On the third day after the Ripley release, Witteman got more bad news. "We were sitting in the road, and we could see a bird. We could see the color band, we could see the radio, the antenna coming off the radio. But we could not get a signal, even pointing the antenna right at it. That was discouraging." The radios appeared to be failing.

Three days later one of the rails was found dead on the sabana road, run over by a vehicle. During the next week, ten more birds were found dead.

On February 15, Witteman and Taisican were driving up to the sabana. Witteman turned on the radio-tracking equipment as they passed a ranch and suddenly got a signal.

"I thought, 'This can't be right' . . . So, we pull off the road and I get out and I point the antenna at this guy's ranch and I'm still getting a signal," Witteman said. "So, we go knock on the door, and the guy has a big chicken pen with a pigpen in the middle and there in the pigpen is one of the rails." The poor bird was being nipped and trampled by the pigs.

As best as Witteman could surmise, the rail had walked all the way down the road from the sabana, seen the chickens and chicken feed, and decided to help itself. But it had gotten caught under the corner of the pen while trying to get away from the pigs.

Somebody would have to go into the mucky pen and rescue the bird. Witteman and Taisican looked at each other. "Hey, it's your rail,' " said Taisican, who was still smarting from his day in the limestone pit. Witteman dutifully climbed in and wrestled with the pigs for the bird. "By that point the bird is covered with pig shit," Witteman said, "I'm covered with pig shit, and Stan is laughing.

"We got in the car, took it to the apartment, washed it up real quick, and called the vet down in Guam and said, 'This bird is not in a good way.' We tried tube feeding, but it died. Birds are sort of binary. They are either very alive and well or they are dead. There were just too many internal injuries."

Another bird arrived at Witteman's door in the back of a pickup truck. Hardy Richards, an instructor at Rota Community College, had seen the bird walking down the road. Stopped. Picked it up and put it in a Tupperware container and drove over to Witteman's place.

"He drives up saying, 'I've got one of your birds,' " Witteman said. But Richards had placed the snap-top cover on the Tupperware bowl and let it bake in the back of the truck. By the time he arrived at the Senior Apartments, the bird was dead. "We put it down as 'death by Tupperware,' " Witteman said. It is unclear where Tupperware and pig dung were accounted for in Pimm's elegant calculations, but they were certainly taking their toll on Rota.

Ten days after the February release, Witteman had stopped his morning road check with a telescope, for he no longer saw any rails. In another ten days he no longer got any radio signals. Either the devices had stopped transmitting or fallen off, or the birds had died.

By March the tally ran as follows: thirteen rail carcasses had been found. Two radios without birds were recovered, and all the other birds were missing. In one case, Witteman said, "we could hear the radio dying, but we could not get to the bird before it went out. It was discouraging because the radios were supposed to last twelve to eighteen months."

Despite problems, Witteman had gleaned one valuable lesson. "The importance of the radio tracking was that we learned that without it we'd never know what happened to them. . . . After three days we never *saw* another bird."

The rails were having problems because the vegetation on the sabana was so thick that the birds were being channeled onto the roads, where they were being hit by autos and exposed to predators, like feral cats, and finding their way onto ranches and into Tupperware bowls.

Originally, the release was to have occurred during the rainy season, when it was thought the birds would have the best chance of adapting. But the delays had been so numerous and the money so short that the release had finally taken place in the midst of the dry season. In large part, the sabana had been chosen as the release site because it was the wettest place on Rota, year round.

"I've tried to figure out why this site was so unsuccessful when the birds were living in similar areas up at Andersen [Air Force Base]," said Beck, who along with Derrickson chose the actual site. "I remember very well the day Scott and I went up to the sabana. It was the last place we looked at. It was an absolute downpour up there, but the rest of the island was dry. . . . That's what really struck us. 'Whoa, this is just what we want.'

"But what we didn't realize was that the area was at the end of a five-year drought period, and by the time the rails were released the vegetation was much thicker and much higher."

Nevertheless, Witteman, Beck, and Pimm were not discouraged. "The causes of death, in nearly half the cases, were things we could correct. They were being hit by cars, torn up by feral cats," Witteman explained. A more remote and secure site could give the rails a better chance for survival.

It seems that the project would have to move anyway, for the hungry Japanese developers had also found Rota. One day, Witteman turned up at the rail release site to find surveyors' stakes plunked down along the path. After making inquiries, he learned that a Japanese group was planning to build a golf course. Under CNMI law, only Chamorros can own land. But developers had gotten around this on Saipan by taking fifty-five-year leases and having local partners. In this case, Witteman heard, the mayor's son was involved in the project. "I said, 'Well, it looks like I just lost a study area.' "

Soon a sign sprouted on the sabana:

ROTA PLUMMERIA COUNTRY CLUB. PROJECT: EIGHTEEN-
HOLE GOLF COURSE, CLUBHOUSE, 120-UNIT HOTEL.
DEVELOPER: ROTA SOUTHERN CROSS RESORT.

Two other large resorts, with the obligatory golf courses and condos, were soon proposed along the coast by Saipanese and Japanese development groups. Suddenly Rota was looking at projects worth more than $850 million that would cover roughly 13 percent of the island.

From the developers' viewpoint, the deals made a lot of

sense. "The problem is that Guam and Saipan are overdeveloped. . . . Land values have soared and the beachfront property is almost gone," explained Edward Yokeno, a Japanese businessman on Guam who watched the Rota plans with admiration.

"So where do [developers] go? There are only two islands left—Tinian and Rota. Tinian is flat and uninteresting, but Rota is a luscious island, with beautiful beaches and beautiful water. In five or seven years, there will be direct flights from Tokyo to Rota."

For the Rotanese, the developers were just another proof that living was easy. They had been showered with U.S. federal funds and programs and now the Japanese were back offering incredible sums for fifty-five-year leases. Overnight big-wheeled pickup trucks, color televisions, and boom boxes seemed to pop up all over the island.

But the land rush sorely tested some long-cherished tenets of the islanders. "It is a Chamorro belief that you hold on to your land. It is your security. Your children stay with you when you have your land," explained Gerald Calvo, a native and director of the island's program for the elderly.

"But now people are selling their land. All they can think about is the things they can buy with the money. The children are fighting over their parents' land. Some of the elderly are being pressured by their children to sign leases."

How Rota would cope with such a land boom even its leaders couldn't imagine. Mayor Manglona, who was not beyond negotiating with developers, pointed out that there were not enough people on the island to build or staff the resorts or even enough places to house guest workers. "We have no crime, no drugs, but there aren't enough people on Rota to staff big hotels. If they import labor like other islands what will happen? Will these things come also?"

Witteman had two jobs now. One was to do a postmortem on the first release. The second was to find another site. The young graduate student spent the next few months thoroughly searching the island with his tracking equipment in the hope of finding any radios. "I got within 200 meters of every location on the island, so I can clearly say at that point there were no more functioning radios out there," he said.

He then made a quick sprint back to Tennessee to teach a field course in ornithology, which would provide him with $3,000 to continue his work. Then back to Rota and his next assignment—the preparation of a new release site.

Saguagahga Point is a sharp spit of cliffs jutting out into the Pacific on the southwestern side of the island. Below the cliffs, where the ocean breaks on a craggy shore rimmed with a virtually inaccessible forest, is a huge seabird colony.

The land atop the cliffs is rocky and covered with slender but tough *panao* trees that send nets of roots spilling over the hard ground. The trees, growing ten or twelve feet tall, create a thick, low, solid canopy of leaves. The cover is so dense and the soil so poor that there is no understory—no bushes, no grass. Instead the trees create tunnels that go in all directions along the cliffs.

There is little water here. But it is more remote than the sabana—there are no farms on the scrabbly land and only one nearly impassable dirt road to the site—and the rails could easily run through the *panao* maze.

Witteman had studied and scouted the area, repeating his rat and lizard trapping. He had also taken a rifle and done some "feral cat control." Now, all was ready for another release.

After the fiasco of the radios, Witteman had been in correspondence with the manufacturer. Embarrassed by the equipment failure, the company had revamped the radios and sent them back to Rota.

After several delays, including one for Super Typhoon Russ, which careened through Guam and Rota with winds of nearly 150 miles an hour, a second release was scheduled for February 4, 1991.

This time the party was composed of only Witteman, Beck, Muna, and Taisican. They carried the long, coffinlike cages that contained the rails to the forest. The radios had been strapped on days in advance to allow the birds to get used to them—another refinement in technique.

The doors along the sides of the long boxes were open and

the rails shot out. Thirty-two birds were freed. This time radios were strapped to just the sixteen males, in the hope that this would be an efficient way of keeping track of mating pairs.

Once again, Witteman began his radio tracking. At first things looked promising. "The radios actually worked. We got sightings of birds up to fifteen days after the release. Birds had much better site fidelity; none of them ran up the road and out of the area. And the birds were calling to each other," Witteman said.

"I got three or four possible cases of vocalizations. In the first release we didn't get any vocalizations. We went out there and played rail calls and what the rails were supposed to do was chorus back. This time I walked into the jungle and heard chorusing. . . . I found lot of little [rail] shits, which is good, and I found only one dead rail . . . a female."

The birds seemed to be contacting each other, feeding, and staying in the general area. And then the project ran into technical difficulties once again. The "new, improved" radios were attached to the birds by two plastic straps that came out of the transmitter and looped across the birds' chests. The straps were held on by solder, but the solder wasn't holding.

There were sixteen radios and, according to Witteman, "All sixteen of them broke right where the strap was potted to the radio. Sixteen birds lost. That's pretty bad. The entire strap was perfect. They all broke in the same place. A pretty severe design flaw. All the radios beeped perfectly. The radios didn't fail, the straps failed. It was Murphy's law of radio tracking."

It was impossible to determine whether the release was a success. Only two birds, one male and one female, were found dead. But without the radios there was no way of telling what was going on with the other birds.

Witteman was out of money once again. Foreseeing the perils of the project, Pimm had arranged some backup field work for his graduate student's Ph.D. thesis. Witteman now left Rota to deal with the remainder of his dissertation.

Beck, however, was not dissatisfied, with the result. "We didn't run into any problems that showed us that establishing the rail on Rota would be impossible. We just have to be able to release more birds, more often," he concluded. But for the moment there was no money, no personnel, and most of the birds were still thousands of miles away.

Pimm had calculated that the chances of a release succeeding were small, somewhere between 5 percent and 15 percent. If nothing else, the rail release program on Rota had proved him right on that score.

21

Spring 1990

he brown tree snake's attacks on puppies and other newborn animals had given David Chiszar an idea. The notion had been lurking on the fringes of his thoughts awhile, but now the assaults had swung the spotlight of consideration full upon it. He had not come to the idea haphazardly. For almost five years now, Chiszar, a professor of animal behavior, had been wrestling—intellectually—with the brown tree snake. But while Fritts and Rodda trekked through Guam's forests in search of their answers, Chiszar's hunt was conducted in a windowless, sub-basement laboratory at the University of Colorado.

It was there that he had placed the snake in one controlled test situation after another, only to be repeatedly confounded by the animal. Still, "by turns," as he described it, he had learned much about the brown tree snake during these five years. Now, he had one more theory, one more challenge for the snake. But this experiment and his very first, which had been a complete failure, were unalterably linked, and so perhaps the best way to understand the new idea was to start at the beginning.

Soon after returning from Guam in 1985, Tom Fritts telephoned Chiszar. "I called him up because he is one of the world's hotshots on how snakes trail, attack, and kill prey. He is a 'sensory cue man' and I was interested in the sensory perception of the snakes. What attracts them? What turns them on to feeding? Was there something that I could use to make my work in controlling them easier?" Chiszar was also a likely choice since he and Fritts shared the same herpetological father figure—Hobart Smith, one of the grand old men in the field and the foremost authority on Mexican reptiles and amphibians.

Fritts called Chiszar at his university lab in Boulder, Colorado. "Hey, sounds great. I've always wanted to work on colubrids. All I've got around here are cobras and rattlesnakes," Fritts remembers Chiszar replying. The part about rattlesnakes and cobras was certainly true. Chiszar had devoted twenty years to analyzing the hunting of rattlesnakes and had written such articles as "Predatory Behavior of 'Nervous' Rattlesnakes" and "Strike-Induced Chemosensory Searching: Do Rattlesnakes Make One Decision or Two?"

The kind of work that Chiszar did was very different from the field biology practiced by Fritts, Rodda, and Savidge. Chiszar's science was more akin to the traditional laboratory brand. He abstracted the real world, simulated it, and then tried to vary a single element and measure change. Still, his point of departure remained the real world, which Fritts and Savidge were trying to describe.

"Unless you design your observation arena so that it is ecologically appropriate for your animal, you are unlikely to see anything resembling the full range of your animal's potentialities," Chiszar explained. "Consequently, if you are brought up in that tradition of thinking, one of the first things you do when you set out to study any organism for any purpose whatsoever is give a little attention to the setting in which you are going to observe the animal."

Chiszar's ability to translate ecological reality into experimental designs led to some Mr. Wizard–like exercises. For example, when a rattlesnake bites its prey, it injects venom and then lets go to avoid a fight. The snake is perfectly content to let its dinner trot off and to wait for the venom to do its deadly work.

The snake then uses chemical cues to find its now prepared meal. But are those cues specific to the envenomated animal?

Can a rattler differentiate between a mouse it bit and any old mouse? In an effort to answer that question, Chiszar misted euthanized mice with musky Halston perfume or a floral scent like Jungle Gardenia. He then allowed the snake to strike one of the mouse carcasses. Two perfume trails were then presented to the animal. With great regularity, the animal followed the perfume trail that matched the mouse it had envenomated.

When Chiszar wanted to find out if this signal was transmitted through the fangs or the membranes in the mouth, he slipped the perfumed mouse carcass into a condom and let the snake bite through it, so that only the fangs penetrated. The snake still followed the proper trail.

But even though rattlesnakes are Chiszar's primary passion, for good measure and for variety he has also studied jumping vipers, cobras, cottonmouths, and death adders. Walking into his lab is an unnerving experience. As one steps through the door and into the darkened room, a sudden hissing—as if steam were escaping from dozens of pipes—is set off. When the light is turned on, the source of the noise is revealed—forty-six rattlesnakes madly shaking their tails. In addition to the four different species of rattlesnakes, there are more than fifty other snakes, deadly ones in the main, like cobras and cottonmouths, ranged in row after row of glass tanks.

Working with such animals got Chiszar's intellectual juices going and made it "enjoyable to put the key in the tumbler at the lab each morning." A refrigerator by the lab door held a variety of snakebite antidotes, just in case someone received a dose of something more than intellectual juices.

"One woman I was involved with for a time accused me of having unresolved Oedipal issues. She took a dim view of my enthusiasm for things that live in holes, in cryptic places, under tree bark—and that bite," said Chiszar, a small dynamo with the street-smart air of his native Perth Amboy, New Jersey, and a taste for string cowboy ties with handworked silver clasps. The latter, no doubt, acquired from years in the deserts of the American Southwest.

Fritts and eight snakes flew from Albuquerque to Boulder. Chiszar assembled some of his graduate students and colleagues for a slide presentation on the brown tree snake. It was an interesting problem, but Chiszar, despite Fritts's recollection, was dubious about taking on another snake project. Fortunately for Fritts, among those attending the seminar were Chiszar's twelve-month-old son, Adam, and his wife, Joan. Adam slept in his stroller. But Joan, a former technician at Emory University's Yerkes Regional Primate Center, was enthralled by the problem. "She urged me to take on the project," Chiszar said. And so, Fritts left his shipment of snakes and returned to New Mexico.

Right from the start, the brown tree snake defied Chiszar's snake wisdom, just as its hamburger-ball eating had astonished John Groves. When he looked at the animal, Chiszar saw its climbing ability, its preference for perches, the rear fangs for grasping prey, the elliptical pupils of a nocturnal animal, and the prehensile tail, and he thought "specialist." This was clearly an animal adapted to a particular set of environmental circumstances and signals, and most likely it had a very select prey base.

In the ongoing struggle for survival, animals tend be either "specialists" or "generalists" in their dining preferences. Humans, for example, who consume many kinds of plants and animals, are supreme generalists. Hummingbirds, supping only on flower nectar, are specialists.

Animals, so the theory goes, evolve into specialists because they develop some advantage that enables them more efficiently to harvest a particular food source. It may limit their menu, but it heightens their success in earning a meal.

For an animal behaviorist there is also a considerable difference between specialists and generalists. "Investigators suspect, with adequate justification, that specialists are likely to be strongly keyed to specific kinds of prey, and, further, to approach those kinds of prey with behavior that is more or less innate and not subject to a great deal of adaptive modification based on experience," Chiszar explained.

"The generalists, on other hand, almost by definition, are animals that are likely to rely a lot on experience and so the theory would predict that since reliance upon learning is likely to be higher in the generalist than the specialist, we are likely to see greater evidence of learning, better development of learning mechanisms, in generalists than in specialists." So specialists should operate by instinct, generalists should depend more on experience.

During the summer of 1985, there were still few hints about the snake's true nature. Fritts and Rodda had not yet seen the animal foraging on the ground, its eclectic menu had not been clearly established, and the snake densities on the island were still only guessed at.

Chiszar started at what he thought was a likely departure point, the chemical cues apt to produce an almost reflex reaction in the snake. It seemed probable that what we would call a sense of smell would be a vital instrument for an animal moving through the rain-forest night in search of prey and so he began by testing to see what odors might arouse the animal.

The standard method used to test the strength of chemical cues on snakes, developed by Gordon Burghardt at the University of Tennessee, is to take a cotton swab, dip it into an extract made of some substance from a potential prey, put the swab within a centimeter or two of the snake's snout, and count the number of tongue flicks made by the animal in a set period of time, usually thirty to sixty seconds. Molecules of odor-causing chemicals adhere to the snake's tongue, and as it is retracted, the tongue touches the Jacobson's organ in the roof of the mouth. The organ in turn is attached to the accessory olfactory nerve.

"If the cotton swab has the right stuff on it, you get not only tongue flicking but predatory attack. *Boom!* It tries to swallow it," Chiszar explained.

Swabs with various rodent aromas—from the skin, from feces, from urine—were offered, but the snakes just stared at the cotton. Perhaps they were spooked by the presence of humans. Chiszar tried the experiments using long forceps. "Nothing would happen, essentially nothing," he said. "They would look at it, maybe a tongue flick or two, and then would either back away into a corner or maybe try to escape. But there was very little tongue flicking and zero predatory attack."

This meant that either chemical cues were irrelevant for the tree snake or there was something wrong in Chiszar's test. He decided that rather than giving up on the chemical cues, he would tinker with his test situation. He tried testing soiled bedding from bird and rodent cages. The material, redolent with prey odors, was placed in a shallow glass plate, called a petri dish. This was paired with a "control" petri dish filled with clean pea gravel or aspen wood shavings.

"The idea was, I could lower the petri dish down into the cage and come back, so they didn't have my looming presence, which isn't a big problem for the garter snake, but might be a big problem for the tree snake. We tried that approach and got the same absolutely dramatic lack of response," Chiszar said.

Then he tried the same kind of experiment using Plexiglas boxes—with lots of holes in the side—filled with various control and odoriferous materials. That didn't work either. "I was absolutely flabbergasted at their behavior," Chiszar said. Somehow Chiszar had expected the brown tree snake and his "big system," the rattlesnake, to have some things in common. They didn't. "All the things I expected them to do," he said, "they did not do!"

Looking back on that summer, Chiszar said: "It was perhaps naive of me to have a bunch of expectations of how the snakes would perform. Nevertheless, I did and where did I get my expectations? Partly from my own previous research, which was mostly with rattlesnakes, and partly from the literature. When you survey the literature, the vast majority of experiments that deal with chemical control of feeding behavior deal with natricine snakes—garter snakes, water snakes, and their allies—and even though I know very, very well that there is a world of evolutionary difference between the natricine snakes and *Boiga*, it is hard to prevent your mind from forming expectations of how this *Boiga* is going to behave based on what you know about other snakes."

Summer had turned into fall. A new academic year had begun and a new graduate student, Karl Kandler from Tübingen, Germany, had arrived at Chiszar's laboratory. One of the first tasks Kandler was given was to run one of the series of Plexiglas box tests on the brown tree snake.

"He also got the same lack of results we got in the previous experiments," Chiszar said. "But Karl was a perceptive young man and he said, 'Ze boweega looks in ze box and zees zere is no prey.' From that observation we distilled the idea that what we needed to do is present the test materials in a context that effectively eliminates vision."

They tried the same experiment in a small, windowless cinder block room—more a large closet than room—with only a dim, 25-watt, red bulb overhead. The light measured only five lux, just barely enough illumination for the human eye to discern shapes and movement. "Karl emerged from those tests," Chiszar recounted, "and said, 'Zee boweega still looks into ze box.'" Chiszar had made his first significant discovery—the brown tree snake could see plainly in the dark.

What Chiszar and Kandler had to do was absolutely eliminate vision as a cue. They could have done that by blinding the snakes, but Chiszar was disinclined to tamper with the anatomy of the animal, believing that introduced a variable he couldn't measure. Instead, they developed an experiment in which two cardboard tubes, twelve inches long and one and a half inches in diameter, were placed in a brown tree snake's cage. One tube held a rat pup and some soiled bedding. A second tube held nothing. The two tubes were placed under the red incandescent bulb. "Karl, being a fastidious German, got the light meter and put it into the bottom of the tubes, and it registered no illuminance under our viewing conditions . . . and that's when we started getting positive results," Chiszar said.

"Eliminate the visual system by rendering it ineffective to the snakes, and by God they switch! *Boiga* are probably a little bit like cobras. They prefer visual information . . . and only when that isn't working properly will they switch [to chemical cues]. But when you create the right situation they switch just fine."

More than 80 percent of the time, the tree snakes in the tests chose the tube containing the rat pup, entered it, and consumed the animal. On top of that, over time, the snakes made their choices more rapidly. The reptiles appeared to be learning as they went along.

As if a code had been broken, Chiszar was now able to devise a host of experiments to test all sorts of aspects of the snake's senses. For example, in order to determine whether the chemical cues the animal used were airborne or residues left on the ground and other surfaces, the floor of another windowless room—sixteen feet long by ten feet wide—was covered with soil and gravel, and a jungle gym for snakes was placed on it. This consisted of three artificial "trees" made out of wooden dowels with ropes strung between the "branches."

In one series of experiments, a snake was placed in the room and then a rat pup in a cylindrical metal "nest" was suspended from the branches. In another, a chemical trail was created by rubbing a rat pup along wood and rope paths. These experiments revealed that while the rodents in "nests" suspended from the trees provoked foraging behavior, the snakes could not easily find the nest. But when a trail was laid, the snake, touching its tongue to the surfaces, made right for the bait.

Chiszar and his assistants also confirmed that the brown tree snake was a mean-spirited animal, for every time they entered the snake's jungle gym room they were attacked. "More often than not, they try to bite you," Chiszar said. "In essence, they are unhappy campers."

Now, with the key stumbling blocks removed from the project, Chiszar found himself with a growing family of tree snakes as the study population rose from the original eight to twenty-six. Soon the brown tree snake took a backseat in Chiszar's lab only to his beloved rattlers.

As he continued to work with the snake, Chiszar made other remarkable discoveries. He started to measure its ability to feed and found that the animal could literally change the shape of its head to accommodate large prey. Normally the head was oval, but it became triangular when presented with big quarry. This enabled the snake to consume prey up to 60 percent of its own body weight. That would be comparable to the average man eating a 100-pound side of beef in a gulp.

Among snakes, vipers have always been considered the animals with the greatest specialization for engulfing large prey, but the brown tree snake put even them to shame. One study, coauthored by John Groves, that compared vipers and non-vipers stated: "The largest prey item successfully engulfed by nonvipers was 18.4 percent of the snake's mass. Vipers swallowed rodents as large as 36.4 percent of their own masses and failed to swallow rodents that were forty and forty-five percent of the snakes' masses."

In Chiszar's experiments a brown tree snake had swallowed a rat 61.7 percent of the snake's weight. In eleven other trials, snakes had consumed rats of 40 percent to 50 percent of their mass.

In collaboration with Hobart Smith and Scott Weinstein, of the Department of Toxinology at the U.S. Army Medical Research Institute, Chiszar also discovered that bigger brown tree snakes produced surprisingly large amounts of venom—more venom than had been found in other rear-fanged colubrids that had been similarly tested. The researchers hypothesized that as the snake grew larger and moved to larger prey, it naturally produced more venom.

The snake's climbing ability also drew Chiszar's attention. In the field, Fritts had already observed the ease with which the snake could climb and its ability to support up to three-quarters of its body. Chiszar took this a step further and in laboratory experiments showed that the snake used a unique form of movement—throwing loops of its body one way and then the other and clamping these loops around even the thinnest branches—that allowed the animal to move very rapidly. Chiszar called this "stenegryous undulation."

Furthermore, on rough surfaces it could use this motion to move vertically. "This means that the brown tree snake is capable of rapid movement over rough surfaces at all angles. Consequently, these snakes would be able to execute rapid pursuit and capture tactics on natural bark, facilitating the capture of birds, as well as geckos and other arboreal vertebrates," Chiszar wrote.

Gordon Rodda had observed an incident in which a dove, seeing a five-foot snake, took wing, but the serpent moved so fast that it closed on the bird before it could fly to safety. "It is probable that the snake was not simply striking from ambush," Chiszar wrote, "but had moved its entire body to reach the bird.

[Stenegryous] undulation makes this kind of rapid pursuit possible, and the snake apparently capitalized on its ability."

When Chiszar had first looked at the snake, he had thought "specialist." He knew now he had been wrong. "I think most herpetologists you might interview would confess that they believe arboreal snakes to be highly specialized creatures," he said. "I approached the brown tree snakes with this point of view in mind, and they turned out to be something other than that. They are much more plastic and flexible than I thought they were going to be. They may be one of the best generalists we have ever studied."

In fact, the brown tree snake would even eat hamburger, dog food, and garbage. There was no record of any snake in the whole world doing that. But when Chiszar tried giving offal to some of his other snakes, lo and behold, he found that his cottonmouth would also eat the wastes. Chiszar was learning not only about brown tree snakes but perhaps about other snakes as well.

"The belief that so many of us have that you must offer just the right prey, presented in just the right way, the right odor, the right thermal cues, comes partly from zoo experience and partly from our childhood experiences of trying to keep animals alive in cages," he said. "But this situation is one where we may have created an overgeneralization, tantamount to a superstition. We have a lot of superstition in science."

Chiszar had learned that the snake could see in the dark, could use chemical senses when necessary, could learn and adapt, could produce significant quantities of venom, could eat a wide variety of foods, and had the capacity to ingest prey almost as big as itself. It could also move more rapidly than a bird could take wing, and come out of the trees and hunt on the ground.

Add to this the fact that the animal could live for long stretches with little food or water and that a female, in all probability, could carry sperm and lay clutches of fertile eggs for years, and the success of the brown tree snake in overrunning Guam became increasingly easy to understand. Its success came not because it was an efficient specialist but because it was an unprecedented generalist.

"The brown tree snake is a colonizing critter. I think there is a big connection between the evolution of generalized predatory behavior and the colonization potential of a species," Chiszar said. The more general the predatory tactics of an animal, the more easily it can adapt to a new setting and new prey and the greater its potential for ecological destruction. Biologists had seen this scenario before with mammals, but never with a reptile. What the brown tree snake had clearly demonstrated was that a reptile could be just as adaptive in its behavior and prey and therefore just as dangerous to an ecosystem. "That's another thing we've gotten from the brown tree snake," Chiszar said.

Chiszar's work had helped to explain why the brown tree snake had been so successful in colonizing Guam and why it posed such a threat to other islands in the Pacific. But he hadn't come up with something to combat the snake, something that Fritts might use to trap the animal. "The problem was that even though in test situations we were able to elicit chemosensory responses, they were never as strong as visual reponses," he explained. But the attacks on puppies and chicks had given Chiszar an idea.

In August 1989, while Greg Witteman was hiking around Rota and getting lost in its forests, Chiszar made his first visit to Guam. He went out into the field to see the snake in its natural Guamanian habitat and got himself tangled in a poacher's snare, proving that an open field can be more dangerous for some people than a laboratory full of poisonous snakes. He also spoke to people up and down the island.

"I heard a number of these stories about baby animals being attacked by snakes, and it got me to thinking," Chiszar said. For a while, he had been considering the need to test a broader array of chemical cues, and now these tales had offered him just the avenue to pursue. Perhaps there was something in the fluids associated with birth that provided a hunting cue to the brown tree snake.

Chiszar went back to the beginning and a new series of cotton swab tests using a variety of substances from pregnant rats, including amniotic fluid and material from the skin and blood. These were used in a series of trials with swabs dipped in tap water, tomato sauce, or perfume as experimental controls.

When Chiszar held out the swab dipped in blood, there was no hesitation. The snake struck. Bingo. He had unlocked another door. The series of experiments indicated an elevated rate of tongue flicking and strike responses for both blood and amniotic fluid from the pregnant mice. But the response was equally strong with blood from male rats and female rats that were not pregnant. It was blood, any blood, that drew the snake.

This discovery presented Chiszar with two new avenues of inquiry. On the practical side, he began collecting commercial blood-based scents used by trappers. Soon, the brown tree snake was being challenged with Russ Carman's Coon Lure No. 2 and Hudson Seal Muskrat No. 1. The snakes also showed a fondness for Johnson's Catfish Candy and didn't seemed to care whether it was red, yellow, or blue.

From a theoretical viewpoint, however, the snake's behavior was once again baffling. Why was it drawn by blood? Chiszar's rattlesnake experiments had demonstrated that the animal picks up sensory cues at the time of the strike and then uses them to trail the envenomated animal. That made eminent sense. But these brown tree snake strikes were not associated with a previous attack. They were simply reacting to the blood. The idea of sharks and tigers, which rip and rend their victims, having an innate blood lust seems natural. But a snake? No one had ever considered it.

Virtually all the documented chemical cues used by snakes involved materials from animal excrement and skin. "This makes good sense because these chemicals are unavoidably deposited by prey and are likely to be encountered by foraging predators," Chiszar was to write in an article on "Mammalian Blood and Snake Predation," and "since snakes do not ordinarily tear or dismember their prey little opportunity exists for blood trails or airborne odoriferous plumes arising from open wounds to serve as cues."

Yet clearly, the blood prompted attacks. "This is," the article pointed out, "the first study to reveal an ophidian [snake] response to internal chemicals of potential prey organisms." Why? No one knew. But Chiszar was determined to find the answer. "From a theoretical viewpoint this opens up a whole new set of questions to be explored. . . . In my opinion, it is the first genuinely new observation since Burghardt started this line of inquiry in the 1960s," he said.

The attacks on children were for Chiszar a variation on this theme. He read a draft of the Fritts-McCoid paper on the attacks and "in a flash of insight" it seemed clear to him that there was also a lot of blood around newborn human babies too. In the spring of 1991, a friend who was an obstetrician provided Chiszar with the placenta from a human birth. The herpetologist placed it in a cage with a brown tree snake to see what would happen. The afterbirth was large and round. Still, with its gaping mouth, the reptile swiftly consumed the entire placenta.

But clearly these attacks were not prompted by the presence of afterbirths. Chiszar obtained some used sanitary napkins and conducted a series of trials, using new pads soaked in water as a control. "*Bam!* They went right after them," he said of the soiled pads. In fact, he obtained some of the very highest predatory strike rates he had ever elicited from the brown tree snake in these tests.

"I remember after my wife gave birth to our kids, she bled for about a month. Now, what happens to those pads? They end up wrapped in some toilet paper and in a wastebasket and then in the trash," Chiszar said.

"What I think may be happening, and for the moment, it is really just a theory, is that the snakes are picking up a cue from menstruating women." Once a snake is drawn to the house, however, something happens. Either the snake, being the observant generalist it is, makes a decision that the baby is the most likely prey and so it attacks or there is an odor attached to the infant that is more locatable than that of a sanitary napkin in a trash container. The only thing that Chiszar is certain of is that "it is clearly feeding behavior" on the snake's part.

And so, in perhaps the most bizarre turn of the bizarre story of the brown tree snake, it is mothers on the island of Guam who may be inadvertently attracting the serpent to their homes, where it then attacks their babies.

October 1990

The weekly flight from Agana, Guam, to Port Moresby, Papua New Guinea, landed at about 11 P.M. on Saturday night. Among the passengers were Gordon Rodda and his wife, Renee Rondeau. The airport was packed with a besotted, noisy crowd. "Saturday is pay day. A lot of people get drunk, go out to the airport, watch the planes come in. Drink some more. Throw bottles. It is an exciting place to land," Rodda said.

The couple worked their way through passport control and passed the big sign warning visitors about endemic malaria. They were supposed to be met by someone from the Australian National University. This was more than a courtesy. There was no place on Guam to change American dollars into Papua New Guinean kinas and only kinas are legal tender on the island. But no one was there to meet them. It was near midnight, Rodda and Rondeau were effectively broke, and their only company was a drunken congregation of airport gawkers.

Port Moresby, the sprawling, shabby, graffiti-riddled capital of the island nation, is not a place to go ambling about in the dark. Gangs of "rascals" roam the city, robbing, raping, and generally terrorizing the place. It is so lawless that the government is forced to impose curfews and the U.S. State Department has issued travel advisories to all Americans venturing onto the island.

"I begged a security guard to let me use the phone and managed to raise somebody and they finally came over and got us. It had all been arranged in great detail beforehand, but of course it didn't work," Rodda said.

Rodda and Rondeau, now both Fish and Wildlife Service biologists, had come to New Guinea and the Solomon Islands in an effort to determine where Guam's brown tree snake had come from and to see what in its native range kept the snake in check. Was there perhaps something as simple as Tom Seibert's *Pareuchaetes* moth that could solve Guam's snake problem?

While he believed that this exploration would be useful, Rodda was dubious that he would find some magic, biological bullet to use against the snake. "When you tell the man on the street about all these snakes on Guam the almost knee-jerk response is 'Well, what eats them in the native range?' The concept of the balance of nature has penetrated into the public consciousness, so the working assumption is that something must eat it and therefore keep its numbers down.

"That comment has not only come to us from the public, which is understandable, but it's been pushed very hard and repeatedly at us by politicans and officials in the Department of Interior, who should know better."

Not everything is eaten by something else. Take human beings. They have no significant predators, and still, for thousands of years, the species has had a relatively stable population.

Setting up this expedition to explore the natural history of the brown tree snake had already been an arduous exercise. "The process of obtaining the permits to do the work from the Solomon Islands and New Guinea governments was very protracted. I found out only when I got there that their general impression seems to be that this process *should* take about eighteen months."

New Guinea is the second largest island in the world. Only Greenland is bigger. It is also one of the least explored corners of the Earth. It was not until 1933 that the interior highlands of the California-size land were first visited by Western explorers. They found primitive tribes brandishing stone weapons and speaking uncharted languages.

Today the island is divided into two political entities. The

western half is Irian Jaya, an Indonesia province. The eastern
half is Papua New Guinea, a former Australian possession and
now an independent state.

Large, tropical, with a remote interior, New Guinea holds a
rich variety of plants and animals, including the brilliantly
plumed birds of paradise, which had drawn Larry Shelton here
seven years earlier.

The exercise that Rodda and Rondeau were about to under-
take was, however, much more complicated than capturing a
single species. Their goal was to attempt to describe the island's
complex ecosystem and the snake's place in it. They wanted to
know what, if anything, ate the snake and what the snake itself
ate. To do that, they would have to know something about the
insects, lizards, birds, and other animals of the island.

But not very much was known about the wildife of New
Guinea. "We had a really serious difficulty that I don't think
most people appreciate. . . . We were going to a place for which
there is not only no field guide to identify animals, there are no
pictures, no keys. You're really on your own," Rodda explained.

For ten months, he diligently searched for whatever information was available. He obtained computer printouts of New
Guinea specimens in museums around the world and went over
them item by item, questioning curators about precisely where
on the island each sample had been found.

He sifted through all the data and records of the 1994
Archibald expedition to New Guinea, again noting the locations where all their specimens had been found. Finally, he
picked the brains of naturalists who had been to New Guinea
and written on the biology of the island. All this he poured into
a computer program designed to create a biogeographical profile of the island.

While Rodda tried to assemble a sketch of New Guinea's fauna,
he and his colleagues were increasingly aware that the snake,
like wind and rain beating on rock, was slowly changing the face
of Guam's ecosystem.

Eight species of forest birds had been wiped off the island.

Three, the Marianas fruit dove, the white-throated ground dove, and the cardinal honeyeater, still existed on other islands. Two, the rail and the Micronesian kingfisher, lived only in captivity. Three more, the flycatcher, the bridled white-eye, and the rufous-fronted fantail, were extinct.

Several other species—including the Marianas crow, the Micronesian starling, and the island swiftlet—had seriously declined in numbers, but still hung on.

It was not only birds that had been affected. Gary Wiles had watched his population of Marianas fruit bats dwindle and still there were no juveniles to be seen in the colony. Shrews and other rodents had disappeared from the forests as well.

There also appeared to be a decline in the number of monitor lizards on the island. Adult monitors were among the snake's few predators, but the babies were vulnerable to the serpent. Indeed, it seemed that some of the larger animals on Guam, like the bat, the crow, and the monitor, had not yet vanished only because they were long-lived. But biologists worried that the populations were aging and would eventually disappear.

Rodda's most recent work had also shown that even lizards had been severely affected by the snake. Two species of geckos had disappeared and a third was severely reduced.

Birds, rats, shrews, lizards, bats, poultry, puppies, the brown tree snake ate them all. "Truly, this is a snake with no class consciousness," Rodda said. DAWR biologists conservatively estimated that the snake had destroyed a native avian population of perhaps 300,000 birds.

The disappearance of so many animals was bound to have a ripple effect. Many of the plant communities on the island depended on birds or bats that ate fruit, nectar, or seeds for seed dispersal and pollination. As the birds disappeared, the ranges and success of these plants would also diminish.

"We have a strong hint that one species of plant, *Lantana camara*, which requires fruit-eating doves to disperse seeds, has already undergone a severe restriction in distribution in the past ten years," McCoid said. "That's just a hint. I think it's the tip of the iceberg. I suspect that many of the other plants on Guam require the assistance of pollinators or seed dispersers to maintain their integrity. But the problem is that we don't have any long-term data on plant distribution. . . . So we can't be certain."

Similarly, McCoid believed that the absence of all the insect-eating birds might have tilted some balance in the island's insect population, but again there were no historical baseline counts to use for a comparison. Baker and Barbehenn might have been curious about Guam's little mammals, but no one had been interested in the island's bugs.

What was undeniable was that the forests were now profoundly silent. All the birds were gone. Only the creak of a tree in the wind, the clacking of a coconut crab scuttling over the rocks, or the buzzing of bees penetrated the resolutely mute landscape. When these noises were absent one could look through the vibrant, sun-flecked woods, with perhaps only a soundless butterfly providing movement, and experience a sort of ecological deafness.

There was one other highly visible change: the woods were now thick with glistening little orb spiders spinning huge diaphanous-webbed walls and domes. It was, in fact, impossible to hike through the woods without becoming matted in the spiders' handiwork. The webs, glinting in the sun, stretched three and four feet across paths, and in one Ritidian forest clearing a dozen of the finger-sized, yellow-striped spiders could be seen constructing a shimmering cupola six feet in diameter, some ten feet above the ground.

No one was sure if there were more spiders or if there were simply no birds flying through the woods to wipe away the webs constantly being spun. Whatever the reason, between the silence and the cobwebs, the rain forests of Guam had taken on the aura of a tomb.

The snafu at the Port Moresby airport turned out to be an augury for Rodda and Rondeau's New Guinea trip. Their plan called for spending two months in New Guinea and about a month in the Solomon Islands. On New Guinea, the two young biologists were to explore Wau in the central highlands, the area near Mandang City on the northern coast, and two offshore islands, Kakar and Manus.

Within a few days Rodda had picked up a giardia parasite and was battling intestinal problems and diarrhea for the rest of his stay. "The really bad spells always hit me traveling one place to the next. When you are strapped into a four-seater plane is not when you want to have an attack of the screaming runs," Rodda said.

As the two researchers went about their work, they found the diligently compiled species list was of limited value. "It was just a list without any information on how to identify species. So, you'd come up with something and say: 'What is it? . . . I don't know.' "

As they worked through the research sites, to their dismay they discovered that the list—in spite of Rodda's rigorous efforts—was off by 50 percent in some cases. Often they found species that weren't on the list. Other times they simply couldn't find species that were on the list and were supposed to be living there.

In an effort to analyze the snake's prey base, Rodda set out sticky traps to sample lizard populations and went out at night with a headlamp to search for lizards, frogs, and snakes. Assessing the avian population was as tricky for the young herpetologist as having to learn about snakes had been for Julie Savidge.

Rodda and Rondeau carefully recorded the varieties of birds in their study. While they did not conduct censuses, it was clear that New Guinea still had viable avian populations and that the loss of habitat rather than the snake was the greatest threat to the birds.

Most of the time the two stayed in huts at native villages, with no running water or electricity. "We had a roof over our heads," Rodda said. "But it is always more challenging to do science when you don't have a way of recharging your headlamp."

Rodda did not discover a single predator that kept the brown tree snake population in check on New Guinea. Rather it appeared that the snake population was lower and the snakes themselves smaller because life was harder for the reptile.

"I think it is safe to say there is a dramatic difference in the availablity of at least the reptile prey," he said. For example, the abundance of geckos was limited. Near the village of Wau, they trapped an average of .2 geckos per hour. On Guam the number was six—thirty times as many. On snake-free islands in the Marianas chain, there were ten captures per hour. The same held true for skinks, which had capture rates on Guam twenty-five times higher than New Guinea.

Ironically, Guam's most abundant skink, *Carlia fusca*, like the brown tree snake, was not native. Both prey and predator had been imported to the Micronesian island.

Without these lizards the small and middling snakes could not survive and as a consequence Rodda determined that survival rates for juvenile brown tree snakes simply weren't as high on New Guinea as on Guam.

As for the birds, unlike Guam's avian population, which was not adapted to coping with the snake, the birds on New Guinea and the reptiles had had a "coevolutionary experience," which probably had allowed the birds to develop defensive behavior.

Something as simple as building nests only on the most fragile of branches, so that the slightest move by an approaching predator sent a shimmer of warning, could be enough to thwart the snake and preserve a species.

"The snake simply doesn't reach the densities we see on Guam. There were no lines we ran where we saw the same number of snakes we see on Guam," Rodda said.

The last stop on the trip was Manus, the largest of the Admiralty Islands, some 200 miles off the New Guinea coast. Like Guam, Manus is part coral deck and part volcanic peak. It is, however, far lusher, ringed by mangrove and covered with Australian pine and coconut. It is roughly three times as large as Guam and even farther off the beaten track.

Still, in and around Lorengau, Manus's principal town, there were lots of little reminders that Americans had been here too—like the roads, the bridges, the plumbing, and the Quonset huts. "It was all old, but all American, built by the military," Rodda said.

Just before the World War II invasion of the Philippines, Manus was the largest naval base in the world. There were eight landing strips and as many as 400 ships in its harbor. "It is claimed that over a million men poured through the Admiralty Islands . . . fighting a war with the most highly developed technical equipment the world had ever seen," Margaret Mead wrote in her study of the Manus islanders, *New Lives for Old.* "The Americans knocked down mountains, blasted channels, smoothed islands for airstrips, tore up miles of bush."

After the war, the island was used as a salvage port for materiel from all over the South Pacific. From Manus much of the equipment was transported by ship 900 miles northeast to Guam's Apra Harbor. There it was broken down or melted. Some was shipped back to the States. Some was sold to Chiang Kai-shek's Nationalist Chinese Army. Some was simply dumped.

It was likely that the ancestors of Guam's brown tree snakes had come from Manus. In fact, it turned out to be more than likely. One of the things Rodda had done was to count the scales of all the specimens he collected, as well as those in collections of half a dozen institutions, from New York City's American Museum of Natural History to the California Academy of Sciences in San Francisco.

The counts, which are the equivalent of fingerprints for a species, were most similar for samples from Manus and those from Guam. In particular, the presence of seventeen scale rows around the body ten rows away from the snake's anal vent distinguished the Manus and Guam snakes from those caught anywhere else in the world.

Yes, this is where it had all started some forty years before. The brown tree snakes' ravaging of Guam had been an obscure eddy in the aftermath of war. It had taken nearly three decades for the problem to surface and it had now been another decade since Julie Savidge had first arrived on Guam to census birds with John Engbring.

The history of the brown tree snake and the birds was almost complete. The question now was, what did the future hold?

Chapter

23

May 1992

On Friday, May 24, Congressman Daniel Akaka of Hawaii rose on the floor of the U.S. House of Representatives. "The United States is experiencing a serious but little-noticed invasion whose costs are astronomical," he told his colleagues. "The armies are larger, numbering in the millions, and the battlefront extends from the East Coast to the borders of Texas and California and stretches all the way to my home state of Hawaii—literally from sea to shining sea. And I fear we may be losing the war.

"I speak of the ongoing invasion of alien pest species. California has been fighting a multimillion-dollar war against the Mediterranean fruit fly for years. Texas has already been stung by African bees. And Customs and Agriculture agents in Miami and Los Angeles can no doubt write volumes on the countless alien plants and animal species they have intercepted and destroyed. But who knows how many more have slipped through?

"Today, Mr. President, I am introducing legislation which seeks to protect Hawaii from one of the most dangerous and costly alien pests. The brown tree snake, which quickly established itself on Guam after World War II, now poses a severe health and environmental threat to Hawaii. I hesitate to even call the brown tree snake a 'pest' since this term misleads people to believe it is no more a nuisance than a housefly or gnat."

Akaka then went on to describe the extirpation of Guam's birds, the other ecological havoc the snake had wrought, the power outages, and the attacks on adults and children.

"Hawaii is already fighting on several fronts to eradicate or control alien pest species, such as the banana poka vine, the clidemia shrub, and the miconia tree, which are literally choking out native plant species," the Hawaiian congressman continued. "And every year an estimated thirty-five new alien species arrive in Hawaii. But the threat of most of these species pales in comparison to that of the brown tree snake."

Akaka's bill directed the Departments of Defense and Agriculture to implement "an aggressive screening of incoming cargo, whether through the use of sniffer dogs, traps, or other measures.

"We must take action now," he concluded. "In the last decade, four brown tree snakes managed to reach Hawaii aboard air cargo. All were caught and killed. So far we've been lucky. But then again, luck is a poor substitute for policy. Unless we act quickly, the brown tree snake may soon be calling Hawaii its new home."

The brown tree snake had truly attained a new measure of notoriety, but it wasn't the first alien invader Congress had dealt with. Just five months earlier, a band of midwestern legislators, lead by Senator John Glenn of Ohio, had pushed through the Nonindigenous Aquatic Nuisance Prevention and Control Act of 1991. The new law, PL 101-646, had been prompted by the coming of the zebra mussel to the North American Great Lakes.

Native to the Caspian, Black, and Azov seas, the inch-long mussel, with yellow and black stripes, is believed to have reached North American inland waters in 1986, when an unidentified ship discharged its ballast tanks in Lake Saint Clair near the port of Detroit. Bilge water was the very same avenue suspected of carrying the weed *Chromolaena odorata* to Guam.

The mussel then spread steadily through Lake Erie, Lake Huron, Lake Ontario, Georgian Bay, and the St. Lawrence River. Each female can produce 30,000 eggs a year and marine biologists have discovered concentrations reaching 700,000 mussels in a cubic yard in Lake Ontario.

Filter feeders that suck microscopic plankton out of the lakes, the mussels are already changing the lake ecosystems. They are competing with native plankton feeders and crowding out native mollusks. They are also removing so much plankton that the lakes are becoming clearer and that in turn is adversely affecting light-sensitive fish, like the walleye.

Lakes are a lot like islands—entities with discrete geographical and biological boundaries—and like islands they are far more vulnerable to invading exotics. The accidental release of the Nile perch in East Africa's Lake Victoria, for example, may result in as many extinctions as all those caused by domestic cats reaching oceanic islands.

It was not, however, the peril to clams and walleyes that moved the the legislators to seek a national program to eradicate the mussel, it was the threat to public water supplies and power plants. The animals were growing on the insides of water intake pipes and literally cutting off plants, factories, and water utilities.

In the summer of 1991, the Ontario community of Lincoln, population 16,300, was forced to impose water restrictions when its reservoir level dropped from thirteen feet to six inches, as the mussels choked the water intake pipe.

Municipal workers used high-pressure hoses to blast the mussels loose, but Lincoln Mayor Raymond Konkle lamented: "They'll come back. I don't think there is any stopping them. It is just a matter of time before they inundate the Great Lakes and other water plants are forced to cut back too."

In fact Great Lakes utilities and municipalities were already spending hundreds of thousands of dollars combating the zebra mussel and the federal government estimated that the animal could cause $5 billion in economic damage by the year 2000. This is what prompted congressional action.

The new act established an interagency task force, led by the Fish and Wildlife Service and the National Oceanic and Atmospheric Administration, charged with creating a program to combat the zebra mussel.

In one of those Byzantine twists Capitol Hill is renowned for, Akaka's call for a program to protect Hawaii from the brown tree snake became section 1209 of the zebra mussel law. A brown tree snake task force was also formed to draw up a control plan, and the snake was designated "an injurious species." There was a nice legal symmetry to that. Now, the rail and the kingfisher were officially "endangered species," and the snake was officially "an injurious species."

Back on Guam the brown tree snake was continuing to baffle the herpetologists. Mike McCoid had discovered that the snake population was under great stress and average body fat weights were dropping. Still, the brown tree snake population did not crash. By 1992, however, the bulk of the snake population on the island were juveniles and that did not make any sense.

In most snake populations the majority of juveniles simply do not survive. That is when the animal's abilities are most limited and it is at its most vulnerable. The result is that mature snakes usually make up 60 percent to 90 percent of a population, and in its native range the brown tree snake was no different, with samples showing 60 percent to 70 percent mature snakes.

But on Guam, the portion of mature snakes had been steadily dropping. In Savidge's 1985 samples it had been 44 percent. By the end of the decade it was 22 percent. "There is no other snake we know with a value that low," Rodda said. "We are seeing some kind of shift from juvenile mortality to adult mortality. . . . Why this is happening, we still don't know." He suspected, however, that it probably had something to do with the fact that the prey preferred by young snakes—skinks and geckos—were still more plentiful on the island than the prey for larger, mature snakes.

The scientific forces studying the snake continued to grow. At David Chiszar's suggestion, Keith Kardong, an evolutionary biologist at Washington State University and expert on snake venom, had been brought aboard to analyze the brown tree snake's toxin, and Robert Mason, a zoology professor at Oregon State University, also began studying the reptile's sex life.

Kardong's initial work—separating the "suite of chemicals" in brown tree snake venom—had led him to the conclusion that, while not as potent, it was "similar to cobra venom." Mason, despite two years of effort, had been unable to get brown tree snakes to mate in captivity, but had done extensive analyses of the animal's hormones.

Earl Campbell, a young graduate student, had also arrived on Guam to work on a "snake exclosure" project that would be his field work for a doctorate from Ohio State University.

Campbell's plan was to design an electrified fence that would keep snakes out of a set of forest patches, ranging in size from two and fifteen acres.

Once the fences are up, he intends to vary the snake densities in each patch by removing snakes with traps or introducing snakes caught elsewhere. "Then I'll try to answer the question of how predator density affects [ecological] communities," he explained.

Tom Fritts had continued his trapping experiments and discovered that an off-the-shelf, store-bought, wire mesh crab trap worked just fine on brown tree snakes and that the blood-based lures Chiszar had come up with were potent attractants for the snakes.

Some experiments were also conducted using fumigants to gas the snakes. But while they did kill the animals, they didn't hold much promise for a comprehensive snake control program, since they were dangerous and expensive.

By the end of this century, it appears that the once obscure brown tree snake will be the best studied, most widely written about snake in the world. That, however, has not solved the problem that gripped Fritts on his very first trip to Guam back in 1984.

Five days after Congressman Akaka sought action to protect Hawaii from the brown tree snake, officials from the Northern Marianas island of Saipan wrote to the federal Fish and Wildlife Service seeking help. Snakes had been found at the island's port, at the airport, and on a road.

"The frequency of reliable snake sightings on Saipan has escalated to the point that all biologists familiar with the situation are convinced that there is an incipient population of brown tree snakes on Saipan," Arnold Palacios, the chief of the CNMI Division of Fish and Wildlife, wrote.

"There have been two sightings in the past week, and five

sightings in the past four months. Even more alarming is the fact that sightings are not restricted to one area of the island. We hope to convince many skeptics here that we have a serious snake problem," Palacios said.

"We know that we must increase local concern and effort," the letter continued, "but I fear that by the time the problem becomes obvious, control and interdiction efforts will be futile."

At the end of the year, the division did its annual Christmas bird census. It appeared that the numbers for some species on Saipan, particularly the smaller forest birds, had dropped 50 percent.

And despite the passage of PL 101-646, on September 3, 1991, two more brown tree snakes were found in the vicinity of Hickam AFB and the Honolulu airport.

The first snake was found at 7:25 A.M. on an airport runway with its head crushed. About five hours later, a ground crew tending a C-5 transport plane at Hickam discovered a second live brown tree snake. Both snakes were more than three feet long.

The night before, a military transport had flown in from Guam and that morning a Continental passenger flight had also arrived from the island. State agricultural officials suspected that the snakes had ridden to Hawaii in the wheel wells of one plane or the other. Perhaps there had been a snake on each flight.

"Finding two snakes in one day is not something to laugh about," said Larry Nakahara, the head of the state Plant Quarantine Division.

"We have come a long way," Fritts said, "but I still don't know if the problem is perceived as truly serious . . . If there is the political will to do something about it."

At Front Royal, Virginia, Scott Derrickson issued an order to all zoos breeding rails to stop. With the release program on hold until 1994, there was no place to put any more birds. The genetic profile of the population still was not ideal. "Some founder lines have bred too much and some not enough," Derrickson explained. But the zoos continued to have problems getting the right birds to mate because of their aggressive nature.

Still, with flip-of-the-switch efficiency the zoos could produce more rails and plans were being made to release more birds on Rota in the future.

The prognosis for the Micronesian kingfishers, however, was problematic. The small, captive population remained beleaguered by disease and cannibalism, and there was still no place in the wild to release birds. The kingfishers were now in the hands of Beth Bahner, the Philadelphia Zoo's animal collections manager. Larry Shelton had left Philadelphia and subsequently retired.

Bob Beck was still on Guam keeping track of both rails and kingfishers. He had become a DAWR supervisor, but had never gotten around to finishing that Ph.D. on vireos. Indeed, he had a new campaign at hand. It was nothing less than a plan to redeem Guam.

Out among scrub forests and cracked, abandoned runways on Andersen AFB, a gleaming new chain link fence runs in a straight line for several hundred yards, takes a neat right-angle turn and keeps going, takes another right and then another. This is the "Area 50" project—an environmental reclamation experiment spearheaded by the United States Air Force.

Yes, the air force. The plan is to restore fifty acres of Guam's woods to as pristine a state as possible. Once the fence is secured, feral pigs and deer, exotics just like the snake, will be eradicated. When Earl Campbell's electric fence is perfected, that will be added and heavy trapping done in an effort to reduce drastically the snake population in the patch. Then the flora and fauna will be closely monitored to see if native species can flourish.

The project was apparently prompted by a heightened environmental concern in the upper echelons of the air force. "Suddenly there was a lot of interest in our putting together some kind of environmental program, and we scrambled around, talked to a lot of people, and stitched this together," said Heidi

Hirsh, a natural resources planner at Andersen AFB.

Under the proposal, which the air force is funding at a level of $150,000, the DAWR, the FWS, and the University of Guam will all participate in the management and monitoring of the forest.

"This signifies a new high-level cooperation between us and the air force," said Beck, "and more importantly it will be the first time we've attempted area-wide snake control. We will be using data that Gordon Rodda has gotten together from his trapping studies, to tell us how intensely we need to trap the area to try to eradicate the snake.

"This is pretty exciting. . . . If it works, maybe we will expand it. And once we think we have the snake under some kind of reasonable control, we are also going to introduce the rail back onto the site.

"Because this area is going to be fenced for snakes, deer, and pigs, we are going to be able for the first time to watch forest regeneration," Beck said. Indeed, those few acres may become the Guam that no one has seen since the Spanish brought deer to the island more than 200 years ago.

Beck and Tino Augon were also excited by the prospect of adding to the "Area 50" project a new electric security device that they had developed to protect the nests of Marianas crows from snakes.

The crows themselves were big enough to fend off the snake, but their nests and eggs had been vulnerable. It had been years since a crow had been fledged on Guam.

Augon had taken on the job of trying to protect the nests. He had first tried using chemicals and slick steel sleeves around the girth of the tall, many-branched *yoga* trees preferred by the birds. Neither worked, and both killed the trees.

Eventually, Augon devised an intricate electrical web, which covered about five feet of the trunk. This proved effective, although he had to add a wire mesh skirt that stuck out about foot to fend off monitors, which also liked crows' eggs. Once a tree was wired, Augon trimmed the neighboring trees to block any arboreal access for the snake.

The design proved effective enough for a crow to be fledged from a nest in 1991. Now Beck is considering wiring trees in "Area 50" not only for crows but for kingfishers.

"If we identify good nesting trees for kingfishers, maybe we will be able to secure them, and in that case we can even get kingfishers back on the island," Beck said.

Perhaps this is the way it has to begin. Making a small island on an island. Not dramatic in the face of the global destruction of rain forests, but a beginning.

Into these fifty acres would be poured the knowledge of Savidge, Fritts, Rodda, Shelton, Derrickson, Pimm, and a dozen other researchers who for more than a decade labored over the plight of Guam's birds and the threat of the brown tree snake. With all this trapping of snakes, wiring of trees, fencing of acres, and reintroducing of captive-bred birds, "Area 50" smacks of the biological management that Pimm predicts is inevitable for the future.

And so just as islands have taught us about evolution and extinction, perhaps now they will also be laboratories for restoration.

"Ironically the limitation of space inherent in island ecosystems (which is at the root of vulnerability of insular forms) can be, and increasingly has been, used to advantage on the resurrection of lost island populations," observed Warren B. King, the former head of the U.S. section of the International Center for Bird Preservation. "Island habitats respond to management more readily than continental ones; in much the same way they can be destroyed more readily."

Fifty acres is not even much of a step in the face of the building boom consuming thousands of acres on the island. In many ways, the conflicts and pressures on Guam—development versus preservation, growing automobile traffic, the pursuit of malls and fast food, the degrading of the island waters, and the overtaxing of utilities—are symbolic of what we all face.

Guam, at least, knows that it is an island with limits. Its people know its boundaries—the place where the land and the ocean meet. But life is filled with boundaries. The membrane that defines a living cell is one boundary. The thin layer of ozone that screens the Earth from the Sun's deadly ultraviolet rays and makes terrestrial life possible is another.

The problem is, we rarely see the boundaries or perceive the limits. Islands, however, never let us forget. They also serve as a test of how well we can conduct ourselves in a world of fixed resources. Granted, it is a small test—a quiz—but it is an important one. "Islands are a link with the land and biota of the past, a measure of how we are conserving our biota at present, and a key to a more varied landscape for mankind for the future," New Zealand biologist I. A. E Atkinson once observed.

The boundary of "Area 50" is not biological. It is fabricated and arbitrary. But it is a line drawn in the sand, so to speak. Here we will try to reclaim and preserve. We must all learn to manage better within the boundaries wherever they are drawn, and fifty acres on Guam is as good a place to begin as any.

Index

About the Author

Mark Jaffe writes on environmental issues for *The Philadelphia Inquirer*. He has been a John S. Knight Fellow at Stanford University and a science writing fellow at the Marine Biological Laboratory in Woods Hole, Massachusetts. He lives with his wife and son in Wynnewood, Pennsylvania.